The Law and Practice Relating to Pollution Control in Italy

There are nine other titles in this series:

The Law and Practice Relating to Pollution Control In

- Belgium and Luxembourg
- Denmark
- France
- Federal Republic of Germany
- Greece
- Ireland
- The Netherlands
- The United Kingdom

The Law and Practice Relating to Pollution Control in the Member States of the European Communities: A Comparative Survey

The series will be updated at regular intervals. For further information, complete the enclosed postcard and send it to:
Graham & Trotman Limited
Sterling House
66 Wilton Road
London SW1V 1DE

All the titles in the series were prepared by

Environmental Resources Limited
79 Baker St, London W1M 1AJ (Tel. 01-486 8277; Tx. 296359 ERL G)

for

The Commission of the European Communities,
Directorate-General Environment, Consumer Protection
and Nuclear Safety, Brussels

The Law and Practice Relating to Pollution Control in Italy

Second Edition

Prepared by

M. Guttieres
International Advocate
President of the International Juridical Organisation

U. Ruffolo
Professor of the Institute of Private Law of the
University of Perugia

for

Environmental Resources Limited

Published by
Graham & Trotman
for
The Commission of the European Communities

Published in 1982 by

Graham & Trotman Limited
Sterling House
66 Wilton Road
London SW1V 1DE

for

The Commission of the European Communities,
Directorate-General Information Market and Innovation,
Luxembourg

EUR 7738

© ECSC, EEC, EAEC, Brussels and Luxembourg, 1982

British Library Cataloguing in Publication Data

Guttieres, Mario
  The law and practice relating to pollution
  control in Italy.—2nd ed.
  1. Pollution—Law and legislation—Italy
  I. Title   II. Ruffolo, Ugo
  III. Environmental Resources Ltd.
  IV. Commission of the European Communities
  344.504'463

ISBN 0-86010-309-9

The views expressed in this publication are those of the authors, and should not be taken as reflecting the opinion of the Commission of the European Communities.

LEGAL NOTICE

Neither the Commission of the European Communities nor any person acting on behalf of the Commission is responsible for the use which might be made of the following information.

All rights reserved. No part of this publication may be reproduced, stored in a retrieval system, or transmitted in any form or by any means, electronic, mechanical, photocopying, recording or otherwise, without the prior permission of the publishers.

Printed in Great Britain by
Robert Hartnoll Limited, Bodmin, Cornwall

Summary List of Contents

1 Introduction	1
2 Air	38
3 Inland Waters and Discharges into Public Sewers	70
4 Coastal Waters and Open Seas	94
5 Deposit of Waste on Land	129
6 Noise and Vibration	141
7 Nuclear Energy	149
8 Control of Products	173
9 The Environmental Problem and the Italian Juridical Organisation	182

Preface

This volume is part of a series prepared in the performance of a contract between the Commission of the European Communities and Environmental Resources Limited (ERL). ERL is a consulting organisation specialising in environmental research, planning and management.

In 1976 a first series was published covering the, then, nine members of the Community. The purpose of those volumes was to explain the law and practice of pollution control in each of the Member States and to provide a summary comparing all the countries in a separate comparative volume.

Since that time many changes in legislation arising from both national and Community-wide initiatives have occurred. ERL was therefore asked to prepare a new series providing an up-to-date review of the law and practice relating to pollution control in the Member States of the European Community.

The series comprises nine volumes concerning the law and practice in the Member States:

Belgium and Luxembourg	Ireland
Denmark	Italy
France	The Netherlands
The Federal Republic of Germany	The United Kingdom
Greece	

and a summary comparative volume.

The aim of this new series, as in the first, is to provide a concise but fully referenced summary of the letter of the law, and a discussion of its implementation and enforcement in practice. Proposals for new legislation which has been drafted but not yet passed are outlined. Where laws have been introduced to comply with Community-wide requirements this is noted.

PREFACE

The publication has two principal objectives:

> to enable the reader to study in outline the provisions in any one Member State; and

> to enable a direct comparison between different Member States.

To facilitate comparison between the national reports, each is indexed following a standard format (the Classified Index) to enable easy reference to the relevant sections of each report.

Presenting a nation's laws accurately in summary form is always a difficult task. There is a danger that, out of context, they may be misunderstood. We have therefore tried to give, in the first section of each report, some of the constitutional, legal and administrative background.

A further danger lies in translation. Although in the English texts we have tried to prepare as accurate a translation as possible, only the authors' original texts in their native languages carry their full authority. These texts are also being published in the individual Member States.

The statement of law in each volume is correct to at least 30 June 1981; in some cases more recent revisions have been included during the period of preparation for publication.

The series will be updated at regular intervals; to receive further details readers should complete the enclosed postcard and send it to the publisher.

ERL would like to acknowledge and express its thanks for the contributions from the national authors and for their cooperation in the preparation of the series.

Finally, ERL also acknowledges the assistance provided by many agencies, which have freely given information and advice, and the help and guidance given by Monsieur Claude Pleinevaux, Mr Grant Lawrence and other members of the Directorate of Environment, Consumer Protection and Nuclear Safety of the Commission of the European Communities.

1982   Environmental Resources Limited
                                                                     London

Detailed List of Contents

Summary List of Contents	v
Preface	vii
Regulations Relating to Environmental Protection	xv
Court Decisions	xxviii
Acts of Parliament Awaiting Approval	xxxi
International Conventions	xxxii
International Conventions with Italian Participation	xxxiii
EEC Directives	xxxiv
Abbreviations	xxxvii

1 Introduction	**1**
1.1 Organisational Structure of Environmental Protection	1
1.1.1 Sources of Legislation	1
1.1.2 Organisation of Public Powers	2
1.1.3 Division of Functions Relating to Environmental Protection	4
1.2 Regulations and Regulatory Bodies on Environmental Matters	8
1.2.1 Constitutional Principles Concerning Environmental Protection	8
1.2.2 Officials and Institutions Concerned with Control of the Territory	9
1.2.2.1 Land ownership and powers of the landowner	9
1.2.2.2 Town planning legislation	10
1.2.2.3 Economic initiative and planning	11
1.2.3 Characteristics of the Repressive Measures Contained in Italian Legislation Against Polluting Activities	12
1.2.3.1 Recourse to provisions contained in the Civil Code	13
1.2.3.1.1 Offences of damage	17
1.2.3.1.2 Crimes of manslaughter and accidental injury	18
1.2.3.1.3 Offences of disaster	18
1.2.3.1.4 Offences punishable as infringements	19
1.2.3.1.5 Waters for human consumption	19
1.2.3.2 Structure of special legislation	19

CONTENTS

1.3 Organising Offices and Associated Bodies Related to the Protection of the Environment   20
   1.3.1 Representatives, Officials and Bodies with Public Functions in Pollution Matters   20
      1.3.1.1 Organs and offices forming part of the central State administration   20
      1.3.1.2 Organs and offices forming part of the peripheral State administration   24
      1.3.1.3 Organisations and offices which are responsible to the regions and other local authorities   27
   1.3.2 Groups with Collective and Widespread Interests and their Importance for the Environment   30

1.4 Protection of the Rights and Interests of Individuals   32
   1.4.1 Interests of Importance in Terms of Environmental Protection   32
   1.4.2 Legal Bodies   34
   1.4.3 Authority, Powers and Rights of the Individual (to Information, to Oppose Authorisation or Permits, to Indictment, to Obtain the Cessation of Harmful Activities or to Compensation)   36

2 Air   38

2.1 General Problems   38
   2.1.1 Introduction   38
   2.1.2 Control by Land-use Planning   41
      2.1.2.1 General   41
      2.1.2.2 Planning procedures under 'anti-smog' legislation   41

2.2 Stationary Sources of Pollution   44
   2.2.1 Industrial Installations   44
   2.2.2 Thermal Installations   46
   2.2.3 Treatment before Discharge and Methods of Discharge   49
      2.2.3.1 Industrial installations   49
      2.2.3.2 Thermal installations   50
      2.2.3.3 Surveillance activities to be carried out by the party responsible for the discharge   51
   2.2.4 Emission Limits and Legal Requirements for Clean Air   52
      2.2.4.1 Industrial Installations   52
      2.2.4.2 Thermal installations   54
      2.2.4.3 Legal requirements for clean air   54
   2.2.5 Control of Raw Materials, Fuels etc.   55
      2.2.5.1 Unrestricted fuels   55
      2.2.5.2 Fuels which can be used freely in Zone A but are restricted to certain installations in Zone B   55
      2.2.5.3 Fuels permitted in both zones only by communal authorisation   56

	2.2.5.4 Fuels which are prohibited for certain uses	56
	2.2.5.5 Controls	58
	2.2.5.6 Special regulations	59
2.2.6	Powers of the Public Authorities	59
	2.2.6.1 Powers deriving from general public health protection legislation (operational over the whole national territory)	59
	2.2.6.2 Powers deriving from anti-smog legislation	60
2.2.7	Rights of the Individual	62

2.3 Mobile Sources of Pollution: Automobiles — 62
   2.3.1 Motor Vehicles: Controls for Approval — 63
   2.3.2 Motor Vehicles: Controls over Maintenance and Use — 65
   2.3.3 Fuel — 66
   2.3.4 Powers of the Public Authorities — 66
   2.3.5 Rights of the Individual — 67

2.4 Ships and Aircraft — 67
   2.4.1 Ships — 67
   2.4.2 Aircraft — 67

2.5 Regional Laws and Other Special Provisions — 68

3 Inland Waters and Discharges into Public Sewers — 70

3.1 Control by Land-use Planning — 71

3.2 Controls over the Use of Water — 73

3.3 Control of Discharges into Watercourses — 74
   3.3.1 Authorisation for the Discharge — 75
   3.3.2 Powers of Inspection, Monitoring and Control — 77
   3.3.3 Duties and Obligations on the Part of the Discharger — 78

3.4 Control over Discharges into Public Sewers — 80

3.5 Emission and Concentration Limits — 81
   3.5.1 Discharges into Surface Watercourses — 81
   3.5.2 Discharges into Public Sewers — 88

3.6 Powers of the Public Authorities — 89

3.7 Rights of the Individual — 92

3.8 Limits and Scope of Complementary Regulations — 92

CONTENTS

4 Coastal Waters and Open Seas   94

4.1 Coastal Waters   94
    4.1.1 Applicable Legislation   94
    4.1.2 Responsible Authorities   97
    4.1.3 Control of Discharges   100
        4.1.3.1 Authorisation for discharges   100
        4.1.3.2 Permitted limits for the discharge   102
    4.1.4 Coastal Activities Relating to Discharges and Coastal Pollution from Fuels   103
    4.1.5 Powers of the Public Authorities and Obligations on the Part of Individuals   104
    4.1.6 Rights of the Individual   106
    4.1.7 Quality Standards for Coastal Waters   106

4.2 Open Seas   108
    4.2.1 Discharges in Open Seas of Substances other than Hydrocarbons   108
        4.2.1.1 State regulations and international conventions   108
        4.2.1.2 Authorisations and controls over discharges   113
        4.2.1.3 Quality and quantity of discharges   115
        4.2.1.4 Powers of the public authorities and obligations on the part of the discharger   115
    4.2.2 Pollution Caused by the Exploitation of the Sea Bed and Offshore Installations   116
        4.2.2.1 State regulations and international conventions   116
        4.2.2.2 Obligations on the part of operators   118
        4.2.2.3 Powers of the public authorities   119
    4.2.3 Pollution from Navigation   119
        4.2.3.1 Discharge or leakage of hydrocarbons from ships   119
        4.2.3.2 Nuclear propelled ships and ships carrying radioactive substances   121
        4.2.3.3 Ships transporting harmful substances other than hydrocarbons   121
        4.2.3.4 Pollution caused by shipping accidents   122
    4.2.4 Contingency Plans in the Event of Accidents Causing Pollution from Hydrocarbons   124
    4.2.5 Civil Liability for Pollution of the Open Sea   126
        4.2.5.1 Pollution caused by shipping   126
        4.2.5.2 Pollution caused by offshore installations   127

5 Deposit of Waste on Land   129

5.1 Control of Deposits or Discharges   131
    5.1.1 Locations for Deposit or Discharge   131
    5.1.2 Method of Deposit or Discharge   132
    5.1.3 Controls over the Quality of the Discharge   134

CONTENTS

5.2 Powers of the Public Authorities	136
5.2.1 Administrative Powers	136
5.2.2 Judicial Powers	136
5.3 Rights of the Individual	137
5.4 New Legislation Proposed for the Disposal of Solid Wastes	137
5.4.1 Municipal Wastes	138
5.4.2 Special Wastes	139
5.4.3 Systems of Disposal	139
5.4.4 Powers of the Public Authorities	140
6 Noise and Vibration	**141**
6.1 Stationary Sources	143
6.1.1 Control over the Siting of Noisy Activities	143
6.1.2 Protection of Workers Against Machine or Plant Noise	143
6.1.3 Emission Limits	144
6.1.4 Powers of the Public Authority	144
6.1.4.1 Administrative powers of the local authority	145
6.1.4.2 Applicable penal provisions	145
6.1.5 Rights of the Individual	145
6.2 Mobile Sources	146
6.2.1 Motor Vehicles	146
6.2.1.1 Controls over design, maintenance and use	146
6.2.1.2 Powers of the public authorities	147
6.2.2 Aircraft	147
6.2.2.1 Controls over design, maintenance and use	147
6.2.2.2 Powers of the public authorities	148
7 Nuclear Energy	**149**
7.1 Introduction	149
7.1.1 Legislation on Nuclear Matters	149
7.1.2 Responsible Authorities	152
7.2 Nuclear Installations	153
7.2.1 Siting of Installations	153
7.2.2 Design and Construction	155
7.2.3 Maintenance and Functioning	156
7.2.4 Radioactive Wastes	158
7.2.5 Obligations on the Part of the Installation Operator	160
7.2.6 Powers of the Public Authorities	160
7.2.7 Rights of the Individual	162
7.3 Radioactive Substances	163

7.3.1	Storage and Use	163
7.3.2	Packaging and Transport	166
7.3.3	Radioactive Wastes	168
7.3.4	Obligations of the Holder of Radioactive Substances	168
	7.3.4.1 Health protection for employees	169
	7.3.4.2 Health protection of the population	169
7.3.5	Legal Standards, Objectives and Guidelines Relating to Levels of Radioactivity in the Environment	170
7.3.6	Rights of the Individual	171
7.3.7	National Energy Plan	171

8 Control of Products — 173

8.1 Synthetic Detergents — 173

8.2 Disinfectants, Insecticides, Pesticides and Similar Products — 176

8.3 Foods — 178

8.4 Mineral Oils — 179

8.5 Toxic Gases — 180

9 The Environmental Problem and the Italian Juridical Organisation — 182

9.1 General Remarks — 182

9.2 Italian Legislation and Community Directives — 192

Bibliography — 197
Classified Index — 203

Regulations Relating to Environmental Protection

1 CONSTITUTION

Art. 9 Const.	(protection of the countryside and artistic and historic heritage of the nation).
Art. 24 Const.	(guarantee of legal protection for the lawful rights and interests of the individual).
Art. 32 Const.	(protection of the right to health).
Art. 41 Const.	(scope of freedom for economic initiative).
Art. 42 Const.	(limits to the protection of property).
Art. 117 Const.	(legislative power of the regions).
Art. 118 Const.	(administrative power of the regions).
Art. 128 Const.	(administrative autonomy of the provinces and communes).

2 CIVIL CODE

Art. 844 c.c.	(prohibition of intolerable emissions on to the property of others).
Art. 866 c.c.	(subjection of lands to hydrological restrictions).
Art. 889 c.c.	(distances for wells, tanks, ditches and pipes).
Art. 890 c.c.	(distances for harmful or dangerous factories or warehouses).
Art. 909 c.c.	(right to the water existing on the property).
Art. 910 c.c.	(use of water bordering or crossing property).
Art. 912 c.c.	(conciliation of conflicting interests regarding water use).
Art. 913 c.c.	(water drainage).
Art. 1171 c.c.	(declaration of new work).

Art. 1172 c.c.	(declaration of feared damage).
Art. 2043 c.c.	(compensation for an unlawful act).
Art. 2049 c.c.	(responsibility of employers and buyers).
Art. 2050 c.c.	(responsibility for dangerous activities).
Art. 2051 c.c.	(responsibility for damage caused to objects in safekeeping).

3 PENAL CODE

Art. 438 c.p.	(epidemic).
Art. 439 c.p.	(poisoning of water or foodstuffs).
Art. 440 c.p.	(adulteration and corruption of foodstuffs).
Art. 441 c.p.	(adulteration and corruption of other objects which damage public health).
Art. 449 c.p.	(culpable acts of damage).
Art. 450 c.p.	(culpable acts of danger).
Art. 451 c.p.	(culpable omission of precautions or defence against disasters or accidents at work).
Art. 452 c.p.	(culpable crimes against public health).
Art. 589 c.p.	(culpable homicide).
Art. 590 c.p.	(culpable personal injury).
Art. 635 c.p.	(damage).
Art. 639 c.p.	(disfiguring or soiling of possessions of others).
Art. 650 c.p.	(non-observance of the provisions of the authorities).
Art. 659 c.p.	(disturbance to the occupation or rest of others).
Art. 660 c.p.	(molestation or disturbance of others).
Art. 674 c.p.	(dangerous depositing of objects).
Art. 675 c.p.	(dangerous placing of objects).

4 CIVIL PROCEDURE CODE

Art. 37 c.p.c.	(defect of jurisdiction).
Art. 41 c.p.c.	(regulation of jurisdiction).
Art. 70 c.p.c.	(intervention in the case by the public ministry).
Art. 75 c.p.c.	(prosecution capacity).
Art. 100 c.p.c.	(interest to act).
Art. 105 c.p.c.	(voluntary intervention).
Art. 106 c.p.c.	(calling of a third party into the trial to whom the case is common).
Art. 107 c.p.c.	(intervention by order of the judge).

5 PENAL PROCEDURE CODE

Art. 1 c.p.p.	(officiality of the penal action).
Art. 2 c.p.p.	(obligation of the statement).
Art. 4 c.p.p.	(report to superior).
Art. 5 c.p.p.	(request for procedure).
Art. 6 c.p.p.	(application for procedure).
Art. 7 c.p.p.	(denouncement).
Art. 9 c.p.p.	(complaint).
Arts. 22, 23 c.p.p.	(civil action in the penal process).
Art. 25 c.p.p.	(relationship between the penal sentence and the civil action).
Art. 27 c.p.p.	(authority of the penal sentence in the verdict of damage).
Art. 28 c.p.p.	(authority of the penal sentence in other civil or administrative judgements).

6 NAVIGATIONAL CODE

Art. 15.
Art. 16.
Art. 18.
Art. 71.
Art. 75.
Art. 76.
Art. 766.
Art. 767.

7 SPECIAL LAWS

(list of principal statutes referred to in the text) (Abbreviations: L = law; DPR = Decree of the President of the Republic; DL = Decree Law; TU = single text; RD = Royal Decree)

L 2248 of 1865	(administrative act).
L 1636 of 3 December 1922	(regulations on nuclear material).
L 2231 of 6 December 1925	(toxic gas).
L 367 of 2 February 1934	(mineral oils).
L 1265 of 27 July 1934	(TU of the health laws).
L 1497 of 29 June 1939	(air pollution).
L 366 of 20 March 1941	(on the collection, transport and disposal of urban solid wastes).
L 1150 of 17 August 1942	(town planning law).

REGULATIONS RELATING TO ENVIRONMENTAL PROTECTION

L 933 of 11 August 1960	(Institution of CNEN (National Committee for Nuclear Energy).
L 238 of 23 February 1961	(protection of coastal waters from pollution).
L 616 of 5 June 1962	(regulations for the safety of navigation and human life at sea).
L 1860 of 31 December 1962	(on the peaceful use of nuclear energy).
L 366 of 5 March 1963	(for the Venetian lagoon and Marano Grado).
L 963 of 14 July 1965	(law on fishing in maritime waters).
L 1103 of 4 August 1965	(regulations on nuclear material).
L 538 of 26 May 1966	(regulations for the safety of navigation and human life at sea).
L 615 of 13 July 1966	(provisions against atmospheric pollution).
L 48 of 27 February 1967	(change of Ministry of Budget to Ministry of Budget and Economic Planning and Institution of CIPE).
L 613 of 21 July 1967	(regulation for research and exploitation of hydrocarbons in territorial waters and on the continental shelf).
L 765 of 6 August 1967	(amending the former).
L 1235 of 19 November 1968	(regulations on the safety of navigation and human life at sea).
L 1008 of 19 November 1969	(radioactive substances).
L 94 of 14 June 1970	(coastal waters).
L 281 of 16 May 1970	(implementation of the regions with ordinary statute).
L 300 of 20 May 1970	(regulations on the protection of the freedom and dignity of the workers, on the freedom of trade unions and their activity in the workplace, and regulations on employment).
L 125 of 3 March 1971	(on the biodegradability of synthetic detergents).
L 437 of 3 June 1971	(for gas discharges arising from vehicles with automatic ignition).
L 863 of 6 October 1971	(economic programming).
L 1083 of 6 December 1971	(safe use of combustible gases).
L 1240 of 15 December 1971	(amending L 933).
L 1102 of 31 December 1971	(mountain communities).
L 15 of 25 February 1972	(date for commencement of functions of regions with ordinary statutes).
L 171 of 16 April 1973	(intervention to safeguard Venice and the Venetian lagoon).
L 443 of 4 June 1973	(pollution caused by exploitation of the sea bed).
L 880 of 18 December 1973	(location of electric power stations).
L 942 of 27 December 1973	(motor vehicles).

REGULATIONS RELATING TO ENVIRONMENTAL PROTECTION

L 109 of 12 February 1974	(peaceful use of nuclear power).
L 341 of 5 June 1974	(coastal waters).
L 382 of 22 July 1975	(with regulations on regional orders on the organisation of public administration).
L 373 of 2 August 1975	(on the siting of nuclear power stations and the production and use of electricity).
L 584 of 11 November 1975	(fixing of no-smoking areas).
L 875 of 19 December 1975	(coastal waters).
L 319 of 10 May 1976	(regulations for the protection of water from pollution).
L 203 of 8 April 1976	(on concessions regarding the planning, construction and management of collection and treatment plants for oily effluent and ballast waters from oil tankers).
L 278 of 8 April 1976	(powers of public authorities).
L 126 of 16 April 1976	(water pollution).
L 690 of 8 October 1976	(conversion of DL 544 of 10 August 1976 into law, amending L 319/1976).
L 10 of 28 January 1977	(regulation for building land).
L 833 of 23 December 1978	(reform of the health service).
L 650 of 24 December 1979	(amendments and integration of L319 of 10 May 1976 and L171 of 16 April 1973).
L 405 of 29 July 1981	(financing of oceanographic research and studies to be conducted in carrying out the Italo-Yugoslav agreement against pollution of the Adriatic Sea).
L 42 of 9 February 1982	(delegation to government of the power to make regulations to carry out directives of the EEC).
L 84 of 5 March 1982	(modification and integration of L 1240 of 15 December 1971 concerning reorganisation of the National Committee for Nuclear Energy).
L 979 of 31 December 1982	(delegation to government of the power to issue norms for implementing some EEC directives).
L 653/1934	(regulations on nuclear material).
L 422/1942	(regulations on nuclear material).
L 1528/1942	(regulations on nuclear material).
L 860/1950	(regulations on nuclear material).
DPR 328 of 15 February 1952	(executive regulation of the navigational code limited to the maritime navigation sector).
DPR 547 of 27 April 1955	(protection of workers from noise).
DPR 854 of 10 June 1955	(toxic gases).

xix

REGULATIONS RELATING TO ENVIRONMENTAL PROTECTION

DPR 303 of 19 March 1956 — (hygiene at work).
DPR 128 of 9 April 1959 — (regulations on the mines and quarries police).
DPR 393 of 15 June 1959 — (TU on road traffic regulations).
DPR 420 of 30 June 1959 — (executive regulation of the above TU).
DPR 185 of 13 February 1964 — (on the safety of installations and health protection for workers and the population against dangers arising from the peaceful use of nuclear energy).
DPR 1704 of 30 December 1965 — (modifying and integrating the above).
DPR 579 of 13 March 1965 — (pollution caused by shipping incidents).
DPR 218 of 6 March 1968 — (new TU on the laws on intervention in the Mezzogiorno).
DPR 1255 of 3 August 1968 — (on phytopharmaceuticals and health protection agents for the preservation of food in storage).
DPR 1639 of 2 October 1968 — (executive regulation of the Law of 14 July 1965).
DPR 1428 of 1968 — (nuclear installations).
DPR 830 of 22 May 1969 — (pollution caused by exploitation of the sea bed).
DPR 1008 of 30 December 1969 — (on the procedure of notification and authorisation for the storage, trade and transport of radioactive material).
DPR 1391 of 22 December 1970 — (executive regulation of the above limited to the thermal plant sector).
DPR 1450 of 30 December 1970 — (regulations for the recognition of ability to operate nuclear installations).
DPR 323 of 22 February 1971 — (executive regulation of L 615 of 1966 limited to the diesel engine sector).
DPR 322 of 15 April 1971 — (executive regulation of L 615 of 1966 limited to the industrial installation sector).
DPR 2 of 14 January 1972 — (transfer to regions with ordinary statutes of administrative functions for mineral waters, springs, quarries, peat bogs and crofts).
DPR 4 of 14 January 1972 — (transfer to regions of functions regarding health assistance).
DPR 8 of 15 January 1972 — (transfer to regions of functions regarding town planning, roads and works).
DPR 11 of 15 January 1972 — (transfer to the regions of functions regarding agriculture, forests, hunting and fishing in inland waters).
DPR 1336 of 30 June 1973 — (pollution caused by exploitation of the sea bed).
DPR 962 of 20 September 1973 — (on protection of the city of Venice and its territory from water pollution).

REGULATIONS RELATING TO ENVIRONMENTAL PROTECTION

DPR 238 of 12 June 1974	(executive regulation of L 125 of 3 March 1971).
DPR 424 of 9 May 1974	(disinfectants).
DPR 519 of 10 May 1975	(on civil responsibility deriving from the peaceful use of nuclear energy).
DPR 992 of 29 May 1976	(protection of the coast).
DPR 1057 of 9 June 1976	(sea waters).
DPR 616 of 24 July 1977	(new transfer of functions to the regions).
DPR 218 of 6 March 1978	(areas and centres of industrial development).
DPR 886 of 24 May 1979	(integration and adjustment of the regulations on mines and quarries police, contained in DPR 128 of 9 April 1959, with the aim of regulating the prospecting, research and exploitation of hydrocarbons in the territorial waters and the continental shelf).
DPR 927 of 24 November 1981	(reception of the directive of the EEC Council no. 831/79 on the classification, packaging and labelling of dangerous substances and preparations).
DM 27 July 1966	(radioactive substances).
DM 19 July 1967	(radioactive substances).
DM 29 October 1967	(research and exploitation).
DM 19 July 1969	(disinfectants).
DM 14 January 1970	(control of insecticides).
DM 28 November 1970	(disinfectants).
DM 15 December 1970	(radioactive substances).
DM 28 December 1970	(disinfectants).
DM 2 February 1971	(direct lines regarding the level of radioactivity in the environment).
DM 14 July 1971	(aquatic produce).
DM 10 August 1971	(biological reserves).
DM 20 October 1971	(disinfectants).
DM 14 December 1971	(aquatic produce).
DM 14 May 1972	(aquatic produce).
DM 19 July 1972	(disinfectants).
DM 7 October 1972	(chemical products).
DM 21 December 1972	(aquatic produce).
DM 1 March 1973	(models for the certificate of fitness of evidence of ability and the personal notebook).
DM 31 July 1973	(control over insecticides).
DM 30 August 1973	(control over aquatic produce).
DM 6 September 1973	(aquatic produce).
DM 23 October 1973	(chemical products).
DM 27 December 1973	(motor vehicles).
DM 7 February 1974	(chemical products).

REGULATIONS RELATING TO ENVIRONMENTAL PROTECTION

DM 1 March 1974 (use of isotopes).
DM 29 March 1974 (aquatic produce).
DM 19 July 1974 (synthetic detergents).
DM 5 August 1974 (reception of the EEC directive on adjustment to scientific progress).
DM 3 August 1974 (chemical products).
DM 14 August 1974 (chemical products).
DM 28 February 1975 (the use of anti-pollution measures at sea).
DM 16 February 1976 (transport of nuclear material).
DM 21 June 1977 (aquatic produce).
DM 8 July 1977 (chemical products).
DM 26 August 1977 (reception of the EEC directive on adjustment to scientific progress).
DM 6 September 1977 (use of anti-pollution measures at sea).
DM 16 December 1977 (aquatic produce).
DM 13 May 1978 (hygiene protection of workers in industry against ionising radiation).
DM 5 May 1979 (reception of the EEC directive on the subject of motorcycle noise).
DM 13 March 1981 (loading, transport by sea, unloading and transhipment of dangerous goods in packaging of the first class (explosives)).
DM 21 May 1981 (packaging and labelling of dangerous substances in implementation of the directive issued by the EEC Council).

Ministry for Public Works (directives on regulation of discharges from public sewers and from civil installations which do not enter public sewers).

RD 2449 of 31 October 1923 (regulations on nuclear material).
RD 383 of 3 March 1934 (atmospheric problems).
RD 1303 of 20 July 1934 (mineral oils).
RD 1265 of 27 July 1934 (location of installations).
TU 148/1915 (atmospheric pollution).
TU 1775/1933 (on water and electrical installations).
TU 393/1959 (motor vehicles).
RDL 3627 of 30 December 1923 (hydrogeological constraints).
DL 376 of 31 March 1947 (administration of the Mercantile Marine).
DL 801 of 30 December 1981 (protection of water from pollution).

8 REGIONAL LAWS

Regulations on internal organisation

BOLZANO LP 29.11.1971 no. 15	institution of the office for the protection of natural resources.
UMBRIA LR 6.9.1972 no. 22	granting to II Advisory Commission of responsibility for the defence of the environment.
BOLZANO LP 19.1.1973 no 6	constitution of Pronvincial Consultative Committee for the protection of natural resources.
SICILY LR 28.6.1973 no. 27	granting to the Health Assessors of responsibility in pollution matters.
ABRUZZI LR 25.7.1973 no. 30	granting to the Regional Health Council of responsibility for protection from pollution.
SARDEGNA LR 1.8.1973 no. 16	institution of the Regional Consultative Committee Against Water Pollution.
TOSCANA LR 6.9.1973 no. 55	granting to the Department of Social Security of responsibility in terms of atmospheric pollution.
TRENTO LP 29.11.1973 no. 59	institution of Provincial Department of Ecology.
TOSCANA LR 27.5.1974 no. 21	purchase of equipment for the sampling of environmental pollution.
PUGLIA LR 17.8.1974 no. 28	granting to the Provincial Consultative Health Committee of responsibility in pollution matters.
LOMBARDIA LR 23.8.1974 no. 49	intervention for the control and prevention of atmospheric and noise pollution.
EMILIA ROMAGNA LR 24.3.1975 no. 19	intervention for the control and prevention of atmospheric and noise pollution.

REGULATIONS RELATING TO ENVIRONMENTAL PROTECTION

LAZIO LR 12.1.1976 no. 2	instition of the local units for social and health services and their responsibility in terms of environmental hygiene and protection from pollution.
BOLZANO LP 26.5.1976 no. 18	institution of Provincial Biological Laboratory with the task of assessing the effects of polluting products on the environment.
SARDEGNA LR 7.1.1977 no. 1	institution of the Assessor for the protection of the environment.
SICILIA LR 18.6.1977 no. 39	regulation for the protection of the environment and for the fight against pollution.

Anti-pollution and financial incentives for businesses

FRIULI VG LR 30.12.1971 no. 70	grant for the interests in industrial funds (businesses must adopt suitable pollution protection measures to qualify).
FRIULI VG LR 4.4.1972 no. 8	reconversion of industrial installations; grants to businesses which adopt measures against pollution and to protect the surrounding environment.
BOLZANO LP 30.8.1972 no. 18	reduction of electricity tariffs for small companies in connection with the observation of the regulations on preventive protection of the environment.
TRENTO LP 2.9.1972 no. 20 and 31.1.1976 no. 12	intervention in favour of small and medium sized industries (only businesses which comply with the regulations on the prevention of pollution can benefit from this aid).

REGULATIONS RELATING TO ENVIRONMENTAL PROTECTION

TRENTO LP 15.12.1972 no. 25	aid for new industrial installations (the beneficiaries must conform with the pollution prevention regulations).
BOLZANO LP 18.12.1972 no. 45 and 22.12.1972 no. 46	concessions of areas and other provisions for the increase of industrial activity; permitted businesses are obliged to observe the regulations on health protection.
FRIULI VG LR 6.1.1973 no. 3	aid to societies of small businesses engaged in protection of health and against pollution.
LAZIO LR 8.2.1974 no. 9	aid to artisan businesses engaged in the protection of the environment.
LOMBARDIA LR 3.4.1974 no. 17	aid to artisan businesses engaged in the protection of the environment.
BASILICATA LR 1.7.1976 no. 22	aid to artisan businesses engaged in the protection of the environment.

Finance and grants for public bodies

LOMBARDIA LR 30.3.1973 no. 22	intervention in favour of local bodies for the constitution of equipped areas.
LIGURIA LR 31.7.1974 no. 22	grants to local bodies for the installation of water purification and treatment plants for solid urban wastes.
PIEMONTE LR 4.6.1975 no. 46	intervention in favour of consortia of local bodies for the disposal of solid wastes.
SICILIA LR 20.4.1976 no. 36	concession of grants to local bodies for the installation of plants for the elimination of residues arising from hothouse cultivation.

REGULATIONS RELATING TO ENVIRONMENTAL PROTECTION

Regulations for the direct or indirect protection of the environment

SARDEGNA LR 20.4.1953 no. 6 and 1.8.1973 no. 16	protection of public waters from pollution.
BOLZANO LP 28.5.1973 no. 17	approval of Varna regulatory plan with indications that the Rio Scaleres should be subject to particular pollution protection.
BOLZANO LP 4.6.1973 no. 12	provisions against air pollution in the open environment, inside buildings and manufacturing establishments.
BOLZANO LP 6.9.1973 no. 61	regulations for protecting the land from pollution.
TRENTO LP 23.11.1973 no. 54	aid for safeguarding and restoring assets of artistic, historic and popular interest.
EMILIA ROMAGNA LR 21.1.1974 no. 3	protection for the environment to allow wildlife to live and reproduce freely.
VENETO LR 31.1.1974 no. 14	grants for purification plants for recycled water.
EMILIA ROMAGNA LR 27.5.1974 no. 20	control of fish breeding in the Comacchio valleys.
TOSCANA LR 4.7.1974 no. 35	protection for the environment to allow wildlife to live and reproduce freely.
LOMBARDIA LR 19.8.1974 no. 48	control of the discharge of waste water.
LOMBARDIA LR 23.8.1974 no. 49	intervention for the control and prevention of atmospheric and noise pollution.
VENETO LR 8.9.1974 no. 49	comprehensive plan and forecast of criteria for the elimination of noxious plants.
TOSCANA LR 9.9.1974 no. 61	control of the indispensable heritage and of public lands for the purpose of land and environmental protection.
PIEMONTE LR 8.11.1974 no. 32	provisions for water purification.
EMILIA ROMAGNA LR 31.1.1975 no. 12	institution of comprehensive committees.

REGULATIONS RELATING TO ENVIRONMENTAL PROTECTION

ABRUZZI LR 9.4.1974 no. 32	development of speleogical tourism and adoption of the provisions against the pollution of water.
EMILIA ROMAGNA LR 19.4.1975 no. 26	regional intervention for the installation of purification plants for recycled waters.
PIEMONTE LR 29.4.1975 no. 23	special aid for water treatment.
TOSCANA LR 5.6.1975 no. 65	institution of the Maremma National Park and prohibition of any activity which could result in alteration or pollution of the environment.
LOMBARDIA LR 14.6.1975 no. 92	control of quarry cultivation.
TRENTO LP 1.9.1975	intervention for the installation of purification plants.
TRENTO LP 11.12.1975 no. 53	minimum standards of water requirements for new installations to safeguard the environment.
VENETO LR 13.1.1976 no. 3	reorganisation of land reclamation consortia and extention of their duties also with regard to plans for environmental protection.
EMILIA ROMAGNA LR 27.4.1976 no. 19	protection of the environment with reference to the restoration of the port system.
VALLE D'AOSTA LR 11.8.1976 no. 34	forecast of provisions for the conservation of the ecological balance for fish life.
PIEMONTE LR 5.12.1977 no. 56	protection and use of land.
PIEMONTE LR 21.8.1978 no. 52	noise pollution.

Court Decisions

CONSTITUTIONAL COURT

Judgement no. 55, 1968	(power to develop the territory).
Bologna Pretura, 18 March 1972	(on art. 844 and art. 32 Const.).
Judgement no. 142, 1972	(town planning and the protection of natural beauty).
Judgement no. 247, 10 April 1974	(on art. 844 c.c.).
Judgement no. 203, 1974	(discharges into the open sea of industrial waste).
Judgement no. 5, 1980	(power to develop the territory).

COUNCIL OF STATE

Judgement of 15 April 1955	(the application of controls over noxious industries to noisy industries).
Judgement no. 253 of 9 March 1973	('Italia nostra').
Section IV, Judgement no. 440 of 29 April 1977	(powers of the mayor).
Section V, Judgement no. 451 of 14 April 1978	(powers of the mayor).

REGIONAL TRIBUNALS

TAR Tuscany, Judgement no. 397 of 13 November 1975	(powers of the mayor).
TAR Lazio, Section II, Judgement no. 177 of 10 March 1976	(health protection).

TAR Lombardy, Judgement no. 10 of 11 January 1978	(powers of the mayor).
TAR Liguria, Judgement no. 207 of 4 May 1978	(health protection).

COURT OF CASSATION

Section II, Judgement of 17 March 1955	(on art. 844 c.c.).
Judgement of 12 July 1961	(nature of riparian rights).
Judgement of 25 March 1969	(penal responsibility).
Judgement no. 2711 of 21 July 1969	(health protection).
Judgement of 20 December 1972	(purity of the atmosphere).
Judgement of 4 May 1973	(atmospheric pollution in non-zoned areas).
Judgement of 6 June 1975	(Art. 659, II comma).
Judgement of 2 July 1975	(Art. 659, II comma).
Judgement of 6 October 1975	(Art 659, II comma).
Judgement of 31 October 1975	(anti-smog measures).
Judgement of 7 November 1975	(Art 659, II comma).
Judgement no. 1976 of 19 May 1976	(health protection).
Judgement of 1 February 1977	(anti-smog measures; responsibility for emissions).
Judgement of 21 February 1977	(penal responsibility for emissions).
SU Judgement no. 1463 of 9 March 1979	(protection of the individual's right to health).
Judgement no. 5172 of 6 October 1979	(protection of the individual's right to health).

LOCAL TRIBUNALS

Tribunal of Turin, Judgement of 9 January 1957	(noise – tolerability limits of).
Tribunal of Cuneo, Judgement of 29 July 1958	(noise – tolerability limits of).
Pretura of Dronero, Judgement of 30 May 1973	(anti-smog measures).
Pretura of Ferrara, Judgement of 31 January 1975	(art. 844 c.c.).
Pretura of Trento, Judgement of 11 December 1975	(anti-smog measures).
Tribunal of Turin, Judgement of 18 March 1976	(anti-smog measures).

COURT DECISIONS

Tribunal of Salerno, Judgement of 6 February 1979, Appeal. Judgement of 4 April 1979   (influx of manufacturing activity in the residential environment).

COUNCIL OF STATE

Opinion no. 582 of 17 October 1972   (air pollution).

Acts of Parliament Awaiting Approval

Senate, Draft Law   Act no. 954 VIII Legislature: Provisions for the disposal of solid waste.

Senate, Draft Law   Act no. 583 VIII Legislature: Provisions for the protection of the sea.

International Conventions

International Convention for the Prevention of Pollution of the Sea by Oil (London, 12 May 1954). Amended 1962, 1969, 1971 (Great Barrier Reef), 1971 (Tanks).
International Convention on the Limitation of Liability of Owners of Ships (October 1957).
Convention on the High Seas (Geneva 1958).
Convention on the Continental Shelf (Geneva 1958).
International Convention for the Safety of Life at Sea (1960 and 1974).
Convention on Third Party Liability in the Field of Nuclear Energy (Paris, 1960).
Supplementary Convention (extends financial limits of Paris Convention) (Brussels, 1963).
European Agreement on the Restriction of the Use of Certain Detergents in Washing and Cleaning Products (Council of Europe, European Treaty Series no. 64).
International Convention Relating to Intervention on the High Seas in Cases of Oil Pollution Casualties (Brussels, 29 November 1969).
International Convention on the Establishment of an International Fund for Compensation for Oil Pollution Damage (1971).
Convention on the Dumping of Wastes and Other Matter at Sea (London, 29 December 1972).
International Convention for the Prevention of Pollution from Ships (London, 2 November 1973).
Convention on the Protection of the Mediterranean Sea against Pollution (Barcelona, 16 February 1976), and its protocols: protocol relating to the prevention of pollution of the Mediterranean Sea caused by discharges from ships and aircraft; protocol relating to cooperation in the fight against pollution of the Mediterranean Sea by hydrocarbons and other harmful substances in the event of emergency.
Convention on Civil Liability for Oil Pollution Damage from Offshore Operations (London, 1976).
Protocol relating to Intervention on the High Seas in Case of Pollution by Substances other than Oil (1973).
International Convention on Civil Liability for Oil Pollution Damage (Brussels, 29 November 1969).

International Conventions with Italian Participation

International Convention for the Prevention of Pollution of the Sea by Oil (*London, 12.5.1954*).
  Ratified by *Law 238 of 23.2.1962*.
Amendment to above *adopted 11.4.1962*.
  Ratified by *Law 94 of 14.1.1970*.
Amendment to above *adopted 21.10.1969*.
  Accepted and implemented by *Law 341 of 5.6.1974*.
Amendments to above *adopted 12 and 15.10.1971*.
  Accepted and implemented by *Law 875 of 19.12.1975*.
Convention on the High Seas (*Geneva, 29.4.1958*).
  Italy participates.
International Convention for the Safety of Life at Sea (SOLAS), *1960*.
  Amendments.
  Approved by *Law 1235 of 19.11.1968*.
European Agreement on the Restriction of the Use of Certain Detergents in Washing and Cleaning Products (*Strasbourg, 1968*).
  Implemented by *DPR 974 of 26.11.1976*.
International Convention Relating to Intervention on the High Seas in Cases of Oil Pollution Casualties (*Brussels, 29.11.1969*).
  Ratified by *Law 185 of 6.4.1977* and *DPR 504 of 27.5.1978*.
International Convention on Civil Liability for Oil Pollution Damage (*Brussels, 29.11.1969*).
  Ratified by *Law 185 of 6.4.1977*.
International Convention on the Establishment of an International Fund for Compensation for Oil Pollution Damage (*Brussels, 18.2.1971*).
  Ratified by *Law 185 of 6.4.1977*.
Convention on the Protection of the Mediterranean Sea from Pollution, with two protocols and relevant annexes (*Barcelona, 16.2.1976*).
  Ratified and implemented by *Law 30 of 25.1.1979*.
International Convention for the Prevention of Pollution from Ships (*London, 2.11.1973*).
  Ratified by *Law 662 of 29.9.1980*.
Protocol relating to Intervention on the High Seas in Cases of Pollution by Substances other than Oil (*London, 2.11.1973*).
  Ratified and implemented by *Law 662 of 29.9.1980*.

EEC Directives

Directive EEC/70/157 of 6.2.1970 regarding permitted noise levels and exhaust devices of motor vehicles (OJ L 42, 23.2.70).
Amendment to Directive EEC/70/157—EEC/73/350 of 7.11.1973 (OJ L 321, 22.11.73).
Amendment EEC/77/212: Council Directive of 8.3.1977, which amends the Directive EEC/70/157 concerning permitted noise levels and exhaust devices of motor vehicles (OJ L 60, 12.3.77).
Council Directive of 20.3.1970 (OJ L 76/1970).
Council Directive of 20.3.70 relating to the provisions to be adopted against atmospheric pollution caused by exhaust gases from automatic ignition vehicles (OJ L 159, 15.6.70).
Directive EEC/72/306 of 2.8.72 for the harmonisation of legislation in member States relating to the measures to be taken against pollution produced by diesel engines used to propel vehicles (OJ L 190, 20.8.72).
Directive EEC/73/404 of 22.11.73 on detergents (OJ L 374, 17.12.73).
Directive EEC/73/405 of 22.11.73 on the method of controlling the biodegradability of anionic surface active agents (OJ L 347, 17.12.73).
Directive EEC/75/440 relating to the quality required for surface waters intended for the abstraction of drinking water in member States (no. 75/440 Council), 16.6.75 (OJ L 194, 25.7.75).
Directive EEC/75/439 of 16.6.75 concerning the disposal of waste oils (OJ L 194, 25.7.75).
Directive EEC/75/442 of 15.7.75 on waste (no. 75/442 Council) (OJ L 194, 25.7.75).
Directive EEC/75/324 of 20.5.75 concerning the harmonisation of the legislation of member States relating to aerosols.
Directive EEC/75/716 of 24.11.75 on the harmonisation of legislation in member States relating to the sulphur content of some liquid composites (gasoils) (OJ L 307, 27.11.75).
Directive EEC/76/160 of 8.12.75 concerning the quality of bathing waters (OJ L 31, 5.2.76).
Directive EEC/76/403 of 5.4.76 relating to the disposal of PCBs and PCTs (OJ L 108, 26.4.76).
Directive EEC/76/464 of 4.5.76 relating to pollution caused by certain dangerous substances discharged into the aquatic environment of the Community (OJ L 129, 18.5.75).

Directive EEC/76/769 of 27.6.76 relating to the harmonisation of the legislation in the member States introducing a restriction on the sale and use of certain dangerous substances and preparations (OJ L 262, 27.9.76).
Council Directive of 23.11.76 on pesticide residues in fruit and vegetables (OJ L 340/76).
Directive EEC/77/312 of 29.3.77 concerning the biological screening of the population against the risk of lead poisoning (OJ L 105, 28.4.77).
Directive EEC/77/538 of 28.6.77 for the harmonisation of legislation in the member States relating to measures to be taken against pollution caused by diesel engined wheeled agricultural or forestry tractors (OJ L 220, 29.8.77).
Directive EEC/78/659 of 18.7.78 on the quality of fresh water which requires protection or improvement to support fish life (OJ L 222, 1978).
Decision EEC/78/176 of 20.2.78 relating to waste arising from the titanium dioxide industry (OJ L 54, 20.2.78).
Directive EEC/78/319 of 20.3.78 relating to toxic and harmful waste (OJ L 84, 31.3.78).
Commission Directive of 30.11.76 (OJ L 32, 1977).
Commission Directive of 14.7.78 (OJ L 223, 1978).
Decision EEC/78/176 of 20.2.78 relating to waste arising from the titanium dioxide industry (OJ L 54, 20.2.78).
Council Decision of 28.5.74 concerning atmospheric pollution caused by combustion engine vehicles (OJ L 159, 1974).
Proposal for a Directive concerning the harmonisation of legislation in the member States relating to the permitted sound level of pneumatic concrete-breakers and jackhammers, presented 31.12.74 (OJ C 82, 14.4.75).
Proposal for the harmonisation of legislation in member States relating to the permitted sound level for current generators for welding, presented 30.12.75 (OJ C 8.3.76), and the permitted sound level for current generators for power supply, presented 30.12.75 (OJ C 54, 8.3.76).
Proposal for the harmonisation of legislation of member States relating to the permitted sound level for tower cranes, presented 30.12.75 (OJ C 54, 8.3.76).
Proposal on the limitation of noise emission by subsonic aircraft, presented 26.4.76 (OJ C 126, 9.6.76).
Proposal for a Directive on construction machinery, 77/113 of 19.12.78.

ABBREVIATIONS

RD	Royal decree
RDL	Law promulgated by Royal decree
TU	Sole text, consolidation act
DM	Ministerial Decree
DPR	Decree of the President of the Republic

1
Introduction

1.1 ORGANISATIONAL STRUCTURE OF ENVIRONMENTAL PROTECTION

1.1.1 Sources of legislation

Italian legislative organisation is characterised by a rigid hierarchy of sources of law. In the event of antinomy (i.e. contradictory regulations), the regulation issued by the higher source takes precedence over the others; where the contradiction arises between regulations issued at a similar level, precedence is given to the most recent regulation.

'Laws, regulations and uses' are sources of law in Italy. A distinction is made between constitutional and ordinary law. As we shall see, legislative authority is also given to the regions.

The solution to each conflict between regulations has to be requested from the judge concerned in the actual controversy over which the conflict arises. However, where such a conflict occurs in relation to constitutional regulations exclusive authority is reserved to the Constitutional Court.

Constitutional laws and ordinary laws are subject to separate systems of legislative formulation (requiring respectively a 'qualified' and a 'simple' parliamentary majority of both Chambers) with the possibility of abrogation by popular referendum. As regards the scope of constitutional law, the Constitutional Charter of 1948 is of great importance; concerning ordinary law, the codes (civil, penal, civil and penal procedure, navigational) are of particular significance. The Civil Code issued in 1942 unifies the previous civil code (1865) and commercial code (1882) and remains faithful to the model of the *Code Napoleon*.

INTRODUCTION

The role of constitutional regulations is not confined to the delineation of the institutional framework and general direction of the activities of the legislator. The basic role of law *in action* of the last decade (with particular reference to the evolution of environmental law) has been the development of the constitutional regulations which can be 'immediately summoned' (i.e. which can be directly invoked and applied, even without having been brought into effect by the ordinary legislation).

Below the Constitution in the hierarchy of sources comes ordinary law which is equal in merit to other regulatory sources (decree laws and delegated legislative decrees in which non-parliamentary but governmental activity plays an essential role in formulation).

The regulatory activity of the Public Administration, central, peripheral and decentralised ('local autonomies' such as the Regions and Communes) is also hierarchically subordinate (regulations, ministerial decrees).

Alongside the State legislative authority comes the regional authority which has exclusive powers in specific matters exclusively determined by the Constitution (article 117 on shared competence) or concurrently or as a result of a delegation contained in a State law (on 'integrative competence') (see 1.1.2 and 1.1.3).

Analogous, limited legislative powers are also attributed to the provinces of Trentino and Alto Aldige.

1.1.2 Organisation of public powers

The public functions regarding general matters are carried out on a dual basis according to a distribution delineated by the Constitution.

On the one hand, there is the State organisation, structured in a series of central and peripheral offices or bodies (some with peripheral ramifications) which form the complex of public administration. In Italian administration (which is an organisation of 'administrative law') great use is made among such offices or bodies of the hierarchical relationship, which creates a pyramid structure culminating in the ministries.

Parallel to the State organisation there is the organisation of minor territorial bodies, endowed with autonomy and giving concrete substance to the territorial organisations of particular communities (article 5 of the Constitution).

The most important of these autonomous territorial bodies is the 'Region' (articles 115 *et. seq.*, Constitution) which is defined as an autonomous body 'endowed with its own powers and functions according to the principles laid down in the Constitution'. The region is in fact endowed with:

Statutory autonomy (there may be one statute, that has to be approved by both chambers of the national parliament).

Legislative autonomy, under which it can issue regional laws which become part of State legislation. The Government can, however, by means of its 'Commissariat' in the region, refuse consent and request new approval by an absolute majority; once this majority has been obtained, the conflict of legality can only be brought before the Constitutional Court, or that of merit before Parliament.

Administrative autonomy, in the sectors of legislative competence or by authorisation from the Government. Controls of legality and merit are provided for by suitable committees.

Financial autonomy, but with its fiscal receipts still largely depending on State tax quotas.

Some regions (with 'special statutes') enjoy a more extensive autonomy as a result of special developmental requirements or those of complex socio-economic situations. These are: Sicily, Sardinia, Trentino, Alto Adige, the Aosta Valley and Friuli Venezia Giulia.

Communes and Provinces are territorial bodies recognised by the Constitution (article 114), but regulated within the structure of State legislation (which is often old, superseded and inadequate for the purposes of autonomy and decentralisation, contained in the Constitution, of which the provincial and communal law of 1915 is an example). In addition, such organisms show signs of the ambiguous nature of autonomous territorial bodies (i.e. with administrative and in part financial autonomy) and of 'areas of national and regional decentralisation' (article 129, Constitution) (i.e. they operate, for certain functions, as peripheral organisations of the State or region).

There is an attempt (albeit only a confused attempt through the lack of rational reform of communal and provincial law) to regain an operational area for the commune (also through the creation of neighbourhood structures such as the 'Councils' of many large cities, which provide greater participation and help to delineate and define the duties of the municipal administration), while the province appears to many as a relic to be replaced by more functional organisations.

The controls of legality and merit (if permitted) over acts carried out by the provinces and communes are today the responsibility of the regions (as is the 'substitutive' control). Controls over the organs of such minor local bodies are still the responsibility of the State.

1.1.3 Division of functions relating to environmental protection

In matters of environmental protection, the first thing to be done is the singling out of the subjects (the State, regions and other territorial bodies) competent in this matter; of the boundaries dividing the various functions; of the way in which these competences figure in the Constitution and in the laws which put these parts of the Constitution into effect.

Environmental and territorial protection does not explicitly figure among the functions constitutionally reserved to regional legislative authority (article 117, Constitution, article 118 for administrative authority). In fact, as we shall see in 1.2.1, the Constitution is not concerned with 'environmental' quality immediately and directly. The protection of the environment results from the importance given to the protection of health, safety, the countryside, etc.

Article 117 of the Constitution does, however, specify certain sectors of intervention which can represent the region's action guidelines for protecting the environment. In fact, the Constitution enables the regional authority to control the following matters:

town planning (probably the most important sector, not only in respect of environmental protection);

the condition of the roads;

aqueducts and public works of regional interest;

navigation and ports on lakes;

mining and peat extraction;

fishing in inland waters;

agriculture and forestry and public health.

Consequently, while on the one hand the State reserves the duty (and thus the power) of directly determining the methods and limits applying to productive activities which are potentially harmful to the environment, on the other hand there is a vast sphere of authority which is now the responsibility of the regions (in respect of which the State can only dictate the legislative framework), and the latter must be considered as the privileged channel for intervention in environmental protection. Furthermore, as we shall see later on, even when exercising legislative authority under State jurisdiction, extensive use is made of the supplementary regional legislative powers requiring the regions to issue detailed regulations in response to local requirements.

The regional statutes also contribute towards the definition of the powers of the region in regard to environmental protection as a competent authority for determining the use and planning of the territory.

The following aspects are covered:

rational organisation of the territory;

protection of the land;

control of waters;

protection of the countryside and of the archaeological and historic heritage;

'purity' of the air and water;

town planning;

environmental protection and prevention and elimination of pollution;

protection and development of indigenous wildlife;

planning of residential areas and manufacturing installations.

Among the aims expressly stated in these regulations are the following:

'to guarantee that the territory will be organised in such a way as to protect the natural heritage, health and general living conditions of present and future generations by promoting a correct relationship between town and countryside';

'to satisfy completely the needs of human beings';

'to eliminate the causes of emigration' (by the enhancement of depressed and mountainous areas);

to encourage 'civil development';

to protect 'health';

to promote the 'social value of the territory by eliminating civil, cultural, economic and social imbalance';

to pursue 'harmonious development and full employment'.

In terms of legislative autonomy, under article 118 of the Constitution the region also exercises administrative functions. Nevertheless, even functions under State authority can be delegated to the regions and performed by them in accordance with State directives.

Thus, in 1972, important functions regarding town planning, sanitation, aqueducts and water supplies were decentralised where public works are concerned. A similar situation exists in respect of health, agriculture

INTRODUCTION

(woodland and forests, land reclamation and improvements) and natural parks. This distribution will be examined in detail in the relevant sections.

Subsequently a full and comprehensive division of State and regional powers in environmental protection matters was achieved by means of DPR 616 of 1977 which transferred functions according to four broad classifications of subject matter, one of which—classified as 'organisation and use of territory'—also includes the 'protection of the environment from pollution'.

On the basis of this DPR:

> the administrative functions of the State and the public corporations are transferred to the regions ... in matters relating to town planning, tramways and motoring routes of regional interest; roads, aqueducts and public works of regional interest; navigation and ports on lakes; hunting and shooting; fishing on inland waters, as relating to the organisation and use of the territory (article 76).

Article 80 of DPR 616 states that

> the administrative functions relating to town planning concern the control of the use of the territory, including all fact-finding, regulatory and management aspects regarding operations to safeguard or improve the land, as well as protection of the environment.

The State continues to be responsible for carrying out coordinating and directing functions, in the identification of 'fundamental lines of territorial organisation', as well as duties connected with the listing of 'seismic zones and consequential building requirements'. Article 81 of DPR 616 also provides

> for the verification that the work to be carried out by the State administration or in areas of State-owned property comply with the town-planning regulations and building requirements to be carried out by the State with the agreement of the region concerned except where works are being carried out for military defence purposes.

The authority of the State over the siting of electricity installations and nuclear power plants, however, remains unchanged.

As regards 'environmental property' referred to in article 82, the administrative functions exercised by central and peripheral State authorities for protection of beauties of nature are delegated to the regions, as far as their identification and protection are concerned, and also the related sanctions.

Aspects covered by this delegated power include the following:

(i) The identification of areas of natural beauty; the Ministry for Cultural and Environmental Affairs is responsible for coordinating the lists of areas of natural beauty approved by the regions.

(ii) The granting of authorisations and approvals for any alterations to these areas.

(iii) The adoption of precautionary provisions, even if the areas such provisions apply to are not included on the relevant lists.

(iv) The adoption of demolition provisions and the imposition of administrative sanctions.

Notifications of considerable public interest concerning areas of natural beauty on the basis of Law 1497 of 29 June 1939 cannot be revoked or amended without prior approval from the National Council for Cultural Affairs.

The Minister for Cultural and Environmental Affairs can prohibit or suspend work which may prejudice the quality of areas of natural beauty, regardless of whether or not such areas are included on the lists.

In matters of 'intervention for the protection of nature', article 83 of DPR 616 transfers administrative functions relating to intervention for the protection of nature, nature reserves and national parks to the regions.

As far as the functions of direction and coordination are concerned, the powers of the Government remain unchanged in the identification of new territories in which to open nature reserves and national parks of an inter-regional character.

The *administrative* activity of the regions is also restricted 'from below' by the existence of minor territorial bodies such as the commune and the province whose organisation, under article 118 of the Constitution, the region should in fact preferably make use of.

Regarding regulatory power, the province (article 241, no. 14 of the Communal and Provincial TU 148, 1915) is given the power to control certain aspects of health; the commune (article 31, no. 6 of the same TU) has the corresponding power for controlling the use of goods or property which are accessible to or usable by the public, and also in matters of hygiene, health and urban and rural policing.

The provinces of Trentino and Alto Adige are exceptions since they directly enjoy legislative authority which includes 'protection of the countryside' (article 5, Constitution Law 1, 10 November 1971), as well as 'health and hygiene' (article 6).

Provinces and communes also enjoy discretional administrative powers to increase or restrict the power of individuals (through authorisations, permits and orders). They also have corresponding powers over administrative policy.

Further powers are vested in the mayor in his role as a Government official (and not as head of the communal administration) in matters of public health and safety.

1.2 REGULATIONS AND REGULATORY BODIES ON ENVIRONMENTAL MATTERS

1.2.1 Constitutional principles concerning environmental protection

It has been said that the Constitution does not mention explicitly the protection of the environment, nor does it specify it as a subject to be protected. Nevertheless, it does set down a series of principles which can be used to delineate a comprehensive provision for environmental protection.

Article 9 of the Constitution needs above all to be considered since it refers to the 'countryside' among other assets of cultural value. This provision appears to refer to the 'countryside' as a particular amenity consisting of places defined as differing in character from the usual habitat of everyday life (conforming to the notion of 'countryside' contained in the legislation in force when the constitutional text was drafted). This definition could also be used in the framework of a comprehensive protection of the health and wholesomeness of the environment.

However, as is widely understood, the reference to 'the countryside' is not sufficient to preclude a search for a basis for constitutional directives for the protection of the environment. Article 32 of the Constitution is more relevant in this respect since it protects health as a 'fundamental right of the individual and in the collective interest'. The principle expressed in clause 2 of article 41 of the Constitution is also relevant since it ensures that economic activity (the Constitution uses the phrase 'economic initiative') cannot be carried out 'in conflict with social benefit, safety, human dignity and freedom'—a principle which provides an immediate precept and is independent of the statutory reservations contained in the article.

The State, public bodies and workers' associations or corporations (article 43, Constitution) also have the powers to intervene in the manufacturing sector *'for the purposes of general benefit'*.

1.2.2 Officials and institutions concerned with control of the territory

It will be opportune here to examine some basic organisational institutions whose form, structure and limitations can affect the protection of the environment.

1.2.2.1 LAND OWNERSHIP AND POWERS OF THE LANDOWNER

The system of land ownership assumes a fundamental role with particular reference to the restriction of powers over land use and the relationship between such powers, public requirements and social benefit. Article 42 of the Constitution recognises and guarantees the rights to private property and in so doing affirms that the law 'determines the methods of acquisition, possession and restrictions in order to ensure its social function'—more than merely recognising ownership, the legislator must have adequate powers to adapt the rights and powers of the owner in accordance with the general interest. In accordance with these principles the legislation on housing property has evolved into an institution which differs considerably from the traditional one. Law 10 of 28 January 1977 on building land has sanctioned this evolution and achieves an almost complete separation between owners and *ius aedificandi* (the right to build). (This, in effect, is a general right to alter the organisation of land: article 1, *loc. cit.*) The owner, solely *qua* owner, does not have the power to change the use of his land without previously obtaining permission to do so (this permission replaces the 'licence') from the mayor, subject to the payment of a 'fee commensurate with the costs of urbanisation and the cost of construction' (article 3). Therefore, through the assumption by the public authority of the power to change land use it is possible to achieve land use planning and to pursue the desired environmental standard. (However, from its judgements no. 55 of 1968 up to no. 5 of 1980 the Constitutional Court has obstructed this aim not a little.)

INTRODUCTION

1.2.2.2 TOWN PLANNING LEGISLATION

The basic provisions for environmental protection, however, do find a meeting point in town planning legislation, defining as they do 'the object of an administrative activity which is clear and distinct in its contents, at the basis of which lies the authority for territorial planning' (GIANNINI).

The sources of this legislation are: Law 1150 of 17 August 1942 (town planning law); Law 765 of 1967 ('bridging law'); Law 1187 of 1968; Laws 291 and 865 of 1971; and, in part, Law 10 of 1977, already mentioned.

Since town planning is one of the subjects included in the list of matters delegated to the regions under article 117 of the Constitution, the provisions of the regional regulations concerning 'planning policy', understood as a basic instrument for rationalising building and activities in the territory, particularly in respect of environmental protection, assume importance. (As an example of the execution of *legislative* functions in this respect, see the Lombardy Regional Law 51 of 15 April 1975 which is exemplary in the coordinated and uniform manner in which it organises the town planning of the regional territory and the measures it provides for the protection of the 'natural heritage and the beauty of the countryside'.)

Town planning is carried out by means of systematic territorial activity at various levels.

General Regulatory Plan: this is the plan for the whole of the communal territory, for which it determines the areas destined for residential expansion, or public services or installations, or manufacturing development. This plan is compulsory for communes included on a list compiled by the region; other communes must formulate a programme of construction under the communal building regulations.

Coordinated Territorial Plan: this concerns areas which are larger than the communes and provides guidelines with which the General Regulatory Plans should conform.

Intercommunal Regulatory Plan; this coordinates the town planning of two or more neighbouring communes.

Special Urban Plans: these are plans concerning the development of given areas as far as manufacturing installations are concerned. Most important are the plans for industrial zones or industrial centres drawn up by zonal or industrial centre consortia in which contractors take part and which are intended for programming the necessary infrastructure and the future zonal arrangements.

The approval of these plans is the responsibility of the regions (since 1972).

Though the possibility of using such instruments for the purposes of environmental protection has been questioned on the basis of a restrictive notion of 'town planning' (as an object of public interest *per se* not connected with the protection of other assets such as health, the countryside, the environment etc.), it has, however, been confirmed by various textual arguments: the town planning law of 1942 requires communal building regulations to contain 'hygiene regulations of particular public interest' (article 33); Law 1187 of 1968 states that General Regulatory Plans must make provision for the relocation of public installations and manufacturing complexes and must also contain restrictions in some zones for historical and environmental reasons and the protection of the countryside: Law 615 of 1966 under article 21 in connection with the preparations of General Town Plans requires industrial zones to be located at a distance from those set aside for housing, also taking into account meteorological factors, to minimise the effect of industrial activities on the residential environment. (In this context see the judgement of the Tribunal at Salerno, 6 February 1979, which maintained that planning authorities and authorities in charge of the protection of the natural heritage should not see themselves as authorities carrying out their functions independently of one another; this judgement was reversed, however, at appeal: App. Salerno, 4 April 1979.)

As regards the Special Urban Plans, particularly those relating to industrial zones, in the absence of specific legal provisions the practice is to take environmental protection requirements into account (rational exploitation of water resources, suitability of certain types of production for the environment, the search for a balanced relationship between territorial expansion and the size of individual installations).

The effectiveness of this urban planning system is seriously hindered by the difficulty of reconciling the requirements of the various territorial entities within a General Coordinated or Intercommunal Plan (which, among other things, requires all the communes involved to adhere to the Intercommunal Plan).

1.2.2.3 ECONOMIC INITIATIVE AND PLANNING

Economic planning can also be used as an instrument to control manufacturing activities which could affect the environmental balance. It is provided for under the Constitution in clause 3 of article 41 ('the law determines suitable programmes and controls in order that public and private economic activity be directed and coordinated in the public interest'), and further defined in article 43 is one of the ways in which the direction and control of manufacturing activities can be achieved (reserving particular economic activities which affect the national economy to the public sector or groups of consumers or users).

Organisations involved in planning are:

    The Ministry of the Budget and Economic Programming.

    CIPE (Interministerial Committee for Economic Programming).

    CICR (Inter-regional Consultative Committee).

    ISPE (Institute of Study for Economic Programming).

    CNEL (National Council for Economy and Labour).

Within the framework of economic planning instruments, Law 836 of 6 October 1971 on the 'Cassa per il Mezzogiorno' appears useful for environmental protection purposes, in spite of the lack of attention shown in applying it, to the aims of protection against pollution.

The law makes all plans for new installations or extensions to existing installations, as well as investment programmes for the Mezzogiorno where particularly important undertakings are concerned, subject to the favourable opinion of CIPE. Among the assessment guidelines there is one concerning the choice of site: the desirability of the investment is compared with the degree of environmental congestion from both the socio-economic and the ecological points of view.

1.2.3 Characteristics of the repressive measures contained in Italian legislation against polluting activities

The response to the requirements created by the increase in pollution in Italy has developed in two ways: on the one hand it has given rise to the appearance of special laws relating to the problem, but frequently with action which is both fragmentary and late; on the other hand, an attempt has been made to avoid loopholes in the legislation by the use or revival of regulations contained in the Civil Code and the Penal Code which have often been used by judges as a temporary substitute for other legislation.

We shall first examine such experiments of an interpretive nature, and then proceed to describe the structural outlines of the special legislation.

1.2.3.1 RECOURSE TO PROVISIONS CONTAINED IN THE CIVIL CODE OR THE PENAL CODE

Starting with the Civil Code, article 84 is of particular importance as it concerns control over the ownership of property. This article provides the owner with the right to take action for the purposes of stopping 'the entry of smoke or heat, exhaust fumes, noise, vibration and similar nuisances arising from neighbouring property' when they exceed the 'normal level of tolerance'. In making his decision, the judge must 'reconcile the requirements of production with the rights of the owner', taking into account amongst other things 'the priority of a particular use'. This article has been invoked to reduce atmospheric and noise pollution, but not water pollution. Its effectiveness in the fight against pollution is limited by various factors.

Action can only be taken by an owner, so the initiative for stopping activities harmful to the general interest rests only with him (attempts at a more extensive or analogous interpretation of the regulation would seem, as we shall see, to have been in vain).

Action therefore too often relates to the protection of ownership to be usefully employed (in the context of 'neighbourly relations') for the protection of health (Cass. 19 May 1976, no. 1976, in *Resp. civ.* 1977, 272, and *Giur. it.* 1978, I, 1, 412) (but the activity of the non-owner perpetrating such nuisance can be passively made legal: Cass. 21 July 1969, no. 2711 in *Riv. giur. ed.* 1970, I, 10).

The regulation is frequently interpreted as symbolising the favour granted by the legislator of the Civil Code issued in 1942 with regard to the property owner: the way in which the judge must adapt 'manufacturing requirements' to 'the rights of property' reveals a preoccupation with not damaging for the sake of property (let alone that of health) the potential profitability of the enterprise.

It is therefore maintained that article 844 of the Civil Code can be revoked by individuals exercising their rights: the owner could permit (and consequently legalise) a nuisance which exceeds the normal tolerable limits on payment of compensation (Cass., Section II, 17 March 1955, in *Giust. civ.* 1955, I, 1093).

As regards the limit of 'normal tolerability', this is interpreted by the courts in a fairly elastic manner, depending on the nature and conditions of the location.

Article 844 of the Civil Code aims in the first place to bring about a cessation of the harmful behaviour. It is then possible to obtain compensation for the damage on the basis of the 'general clause' of responsibility for illicit acts under article 2043 of the Civil Code (as distinct from article 844 of the Civil Code and therefore also actionable by persons other than the owner, e.g. the driver etc.). But it is maintained that, in a case where the judge has established the objective intolerability of the nuisance, but considers the activity not to be illegal because of the over-riding manufacturing requirements, he can replace the right to compensation (which is only applicable in cases of illegality) with an indemnity payment, which does not represent the full amount of the damage sustained, but only part compensation, often representing considerably less than the value of the facilities destroyed by the nuisance. This has been contested with strenuous arguments by those (VISINTINI) who consider nuisance under article 844 of the Civil Code to be a special case of extra-contractual responsibility (with the result of extending effective legitimation) in respect of which the discretion of the judge would operate only to establish whether or not the emissions are tolerable. In this way the reconciliation of 'requirements of production' and 'rights of the owner' is not seen as a source of justification (and therefore of permissibility) for the harmful activity, but as a criterion for further specification of the tolerability of the nuisance.

Attempts at interpreting article 844 of the Civil Code in the most suitable manner with regard to environmental protection have been accompanied by criticism of the article in respect of its constitutional legality, with reference to article 3 of the Constitution (principle of equality) in that it affects the legality of making use of civil protection against nuisance. Reference is also made to article 32 of the Constitution (health protection) regarding the criteria and limits for the effectiveness of the provisions described above (see the ordinance for submission to the Constitutional Court of Pret. Bologna, 18 May 1972, in *Giur. it.* 1973, I, 2, 708).

The Constitutional Court (10 April 1974, no. 247, in *Giur. it.* 1975, I, 1, 586; *Mondo giur.* 1974, 371), while recognising the inadequacy of article 844 of the Civil Code in terms of environmental protection, refrained from declaring it unconstitutional in spite of its defects and limitations, as they can be remedied by recourse to article 2043 of the Civil Code (responsibility for illegal acts) which provides protection for every individual (even the non-owner) and for anything which is prejudicial to health, even if it is caused by nuisance considered 'tolerable' in the light of article 844 of the Civil Code.

Still within the scope of provisions relating to ownership, article 890 of the Civil Code, which is reasonably comprehensive, obliges anyone wishing to 'manufacture furnaces, chimneys, warehouses, animal houses or similar, or to install machinery from which danger or damage could arise' to observe the distances from other properties established by the regulations and (in the absence of appropriate regulations) 'to preserve surrounding property from any damage to its solidity, cleanliness and safety'.

The provision provides for the operation of the communal regulations, but its particular interest lies in the section controlling situations in which such regulations are wanting. In such a case it is the responsibility of the judge to establish whether the construction constitutes a risk of damage and, if so, whether the distance from surrounding properties is suitable for avoiding damage to 'the solidity, cleanliness and safety' of such properties.

As regards water pollution, it has been considered possible to make use of the controls contained in article 909 f.f. of the Civil Code for the use of 'private waters'. This provision prohibits the owner of a property which uses existing water sources to 'divert them in detriment to other land'; it also prohibits the owner of land on a higher level whose water naturally flows into other lower-lying land (which is required to submit to such a natural flow) from 'increasing the flow of water' (and also, according to prevailing opinion, from altering the quality of waters discharged or from causing the presence of extraneous substances).

Articles 2043 *et seq.* of the Civil Code regarding responsibility for illicit acts is also of great importance in respect of damage to the environment is also of great importance because of the characteristics it assumes in Italian legislation. In fact, article 2043 appears to perform the role of a 'general clause' for extra-contractual responsibility and is well suited to dealing with an indeterminate number of cases in point concerning damage, especially where the damage is to health or to other protected rights or interests. The particular criteria for determining responsibility can then be found in the appropriate provisions covering civil responsibility for imputing objective responsibility to the polluter (this is much more appropriate for dealing with responsibility for offences relating to environmental protection than responsibility for negligence). These provisions are:

Article 2050 (dangerous activities) for those *activities* which, as a result of the particular way in which they are carried out or because of technical imperfections relating to safety and control, introduce a risk of damage.

Article 2051 (responsibility of a person for harmful effects for either normal operation or accidents arising from the use of machinery, industrial plant or other articles in his custody).

Article 2049 (responsibility for acts of 'commission') when the damaging act can be ascribed to ('illicit') negligence on the part of the

INTRODUCTION

human element involved in the potentially polluting manufacturing activity.

The United Sections of the Court of Cassation in two recent and important judgements have delineated the basis and protection of the individual situation, with reference to health rights which can be made use of by individuals in the event of damage to the environment (Cass. S.U. 9 March 1979, no. 1463, in *Foro it.* 1979, I, 939; Cass. S.U. 6 October 1979, no. 5172, in *Foro it.* 1979, I, 2302). It is a question of an individual's right to protection being safeguarded by the Constitution in an absolute manner, without restrictions or conditions on the part of other interests, including general and public interests; it is therefore completely and immediately invocable, even as regards the public administration.

Passing on to the use of the provisions dealing with environmental protection contained in the Penal Code (none of which make direct reference to pollution), it should be pointed out that, while on the one hand the threat of penal sanctions can appear the best of deterrents, on the other hand (contrary to the case of civil responsibility) it can become difficult to identify the person responsible for the offence in what are often vast manufacturing organisations, even under article 27 of the Constitution, which states that 'the penal responsibility is personal'. Thus, even without contesting the response of penal law to certain harmful acts, considering its effectiveness as a deterrent, the instrument of (objective) civil responsibility would often seem preferable since it represents a 'cost' for the management.

In broad terms, it can be said that penal responsibility rests with the individual who, as a result of the distribution of work within the business, is placed in a position of considerable organisational responsibility since he is a technically qualified and suitable person for the job (Cass. 25 March 1969, in *Giur. pen.* 1970, II, 433). That is to say that it falls upon the individual who has the power (and financial resources) to provide all the means and technical knowledge suitable for preventing the risk of damage to the environment.

A problem arises when it is a combination of factors which produces a damaging incident, since a polluting situation or deterioration of the environment is rarely traceable to a single factor. But when an appropriate causal connection exists, responsibility can be allotted to the individual in the light of the personal element (negligence or malice), if the perpetrator of the act using normal diligence is able to realise that his action was accompanied by other contributory causes in a way which cumulatively resulted in a damaging incident (in the case of malice, the full knowledge of the perpetrator that the act contributed to a chain of factors producing the damage must be proved).

It is advisable to remember that having obtained an administrative authorisation for carrying out a certain activity does not constitute any extenuation of penal responsibility for that action, nor does it exclude civil responsibility.

The allusion to the administrative activity of the public powers with regard to private activity which may affect the environment leads us to consider a further aspect of the application of the Penal Code: penal responsibility can be identified in public officials who are charged with controlling or regulating private activities (article 328 of the Penal Code).

Let us now consider the different types of offence relating to pollution.

1.2.3.1.1 Offences of damage

Article 635 of the Penal Code punishes offences which cause 'deterioration or make movable or immovable assets belonging to others either totally or partially unusable'. The requirement of ownership by another person of the articles in question (which can also be land) excludes air pollution (air is *res nullius*—the property of no one). However, the article can apply if, following atmospheric pollution, there is subsequent damage to public or private belongings (but in this case it would be more usual to apply article 639 which covers disfigured or soiled articles). It is claimed by some people (CICALA) that not even discharges causing water pollution which are controlled under Law 319 of 10 May 1976 (dealing with protection of waters against pollution, cf. Chapter 3) can give rise to the offence of damage (because of article 26 of the above law which revokes all provisions directly or indirectly controlling discharges into surface waters or groundwaters, and consequent pollution, if they are not covered by this article). Others (AMENDOLA) contest this interpretation (the punishable offence of damage does not regulate the matter of discharges), but even following this opinion the subject of marine pollution remains outside the scope of the provisions under the Penal Code (only offshore anchorages, ports, lagoons and salt water basins connecting with the sea are public property; the sea is *res communis omnium*—common property).

The damage is more serious in some situations under clause 2 of article 635 of the Penal Code, and in particular when it is caused to one of the objects indicated in no. 7 of article 625 of the Penal Code (objects destined for public use, including waters which are public property). In fact, article 625 refers only to personal assets (covering the crime of theft), but the application operating under article 635 of the Penal Code is not limited to personal property since it is an appeal to certain criteria which provide information regarding the aggravating circumstances under no. 7 of article 625 of the Penal Code (also used in regulations relating to damage).

INTRODUCTION

Under Italian legislation the punishable offence of damage is provided for only in the case of malice. It is, however, a question of general or possible malice (knowledge is sufficient, without any intent to damage or *animus nocendi*—intent to harm). The punishable offence of damage coincides with the offences under some special environmental protection laws (see *Pret. Mortara*, 9 November 1977; *Pret. Pavia*, 22 December 1977).

A very delicate problem is the one of establishing what constitutes deterioration or alteration of an object so as to render it unsuitable for the use for which it was intended. Excluding the simple fouling or contamination of water (the TU of the health laws punishes this with a fine only where the water is designed for foodstuffs), the following must be considered of importance for the purposes of the offence of damage: the identity of the party responsible for making the public water unsuitable for becoming the subject of concessions for the uses for which it was intended (food, agriculture, industry) which characterise its 'public utility' or to represent the natural habitat of aquatic wildlife (this is the basis of the nature of the public ownership of those watercourses: Cass. 12 July 1961, in *Foro pad.* 1962, I, 1415).

1.2.3.1.2 Crimes of manslaughter and accidental injury (articles 589 and 590 of the Penal Code)

The effective application of these provisions to matters of pollution encounters a double obstacle. On the one hand there is the problem of causation, since it is not easy to demonstrate that an illness or injury was caused by polluting phenomena, nor precisely to identify the polluting factor responsible among the various causes of environmental deterioration. On the other hand it is difficult to determine which precautions are imposed on the polluter, in the presence of a variety of concurrent factors outside the latter's control.

These problems are only easily overcome when death or injury results from the deterioration of a particular environment for which the business owner is responsible (e.g. the work environment).

1.2.3.1.3 Offences of disaster

Article 434 of the Penal Code contains provision for 'generic disasters' (i.e. not specifically provided for under heading I of title VI of the Penal Code) in which particularly serious acts of pollution may be included. This offence is also provided for in culpable form (a combination of articles 434 and 456 of the Penal Code).

1.2.3.1.4 Offences punishable as infringements

Various criminal situations are required to be considered:

(i) Article 650 of the Penal Code (non-compliance with a provision issued by the authorities for reasons of justice, public safety, public order or hygiene).

(ii) Article 659 of the Penal Code (disturbance of the occupation or peace of individuals), clause 2 of which provides for the case of anyone performing 'a noisy profession or trade in contravention of the dispositions of the law or the directions of the authorities'.

(iii) Article 674 of the Penal Code (punishing those who 'throw or spill things which may offend, soil or annoy people in places open to the public or in private places used by others', as well as those who 'in cases not permitted by law cause the emission of gas, vapour or fumes which may offend, soil or annoy people'); this will be discussed further in connection with article 20 of Law 615 of 1966 in the section relating to air pollution.

(iv) Article 734 of the Penal Code (destruction or adulteration of the natural beauty of places which are subject to special protection).

1.2.3.1.5 Waters for human consumption

Finally, there are cases of offences which are particularly important for the protection of waters intended for use in foods (poisoning or contamination of waters, storage, sale or distribution of poisoned, adulterated or otherwise dangerous waters, etc.—articles 439, 440 and 442–4 of the Penal Code). This aspect will be discussed in the section on water pollution.

1.2.3.2 STRUCTURE OF SPECIAL LEGISLATION

The general principles described up to now coincide with those contained in the special legislation for environmental protection and are still widely applied in spite of the introduction of specific regulations in cases where protection of individuals is required (cf. 1.2.4) or where the crime threshold has been exceeded.

The special laws which include provisions for protection from pollution should therefore be read in conjunction, for aspects that more directly concern the protection of goods and individuals; with provisions in common use such as provisions for closure which can then be applied as an extreme measure of protection when (as a result of violation of provisions in special laws, or even in the absence of such violation) polluting activities regulated by these laws violate the individual rights of third parties or can be regarded as constituting penal illegality.

Another characteristic of the special laws for protection against pollution is that they act as framework laws, normally with respect to regional legislation, but also with respect to implementation regulations, as they are limited to stating the basic principles of the matter in hand, thereby leaving real control to the more technical and adaptable instrument of the regulation, issued by ministerial decree or by decree of the President of the Republic.

1.3 ORGANISING OFFICES AND ASSOCIATED BODIES RELATED TO THE PROTECTION OF THE ENVIRONMENT

1.3.1 Representatives, officials and bodies with public functions in pollution matters

We will now list the organs, offices and bodies holding administrative functions in environmental protection matters, excluding, therefore, the jurisdictional organisations (see 1.4.2). Reference will be made to the points discussed previously under 1.1.2 and 1.1.3 regarding the division of public functions between the State and local organisations under the Constitution.

1.3.1.1 ORGANS AND OFFICES FORMING PART OF THE CENTRAL STATE ADMINISTRATION

Starting with the offices which form part of the central administration of the State, first of all the ministries (the vertex of the administrative organisation of the various sectors referred to) should be examined. To date there is no ministry which unites all of the delegated powers dealing with the protection of the environment to create a coordinated and effective management. Instead, these powers are distributed among:

- the Ministry of Health;

- the Ministry of Public Works (responsibility for public waters, aqueducts, town planning and organisation of the territory);

- the Ministry of Industry (responsibility for exploitation of resources, industrial works, and control of the production and use of hydrocarbons);

the Ministry of Transport and Civil Aviation (primarily of interest for the control of standards for polluting engines and ground and air traffic, and related atmospheric and noise pollution);

the Ministry of the Mercantile Marine (management of the maritime area and its protection against pollution);

the Ministry of Agriculture and Forestry (responsible for the protection of forestry land and for the control of production activity with polluting potential in the agricultural sector);

the Ministry for Cultural and Environmental Affairs;

the Ministry of the Budget and Economic Programming (which determines the methods of development and therefore the provisions for ecological organisation);

the Ministry of the Interior (hierarchically senior to the peripheral organisations for State intervention such as the prefect and the fire service, with functions of environmental defence and control over certain sources of pollution such as heating installations);

the Ministry of Labour (for the protection of safety and quality of the work environment);

the Ministry of Education (for the protection of the countryside);

the Ministry of Scientific Research (for the coordination and financing of studies and research);

the Ministry of Justice (for studies on legislation and the activities of the judges: in this context the Commission on Ecology was formed by this Ministry to carry out an interdisciplinary study of existing legislation and prospects for amendment).

Other important professional bodies operate within the Ministries themselves or in auxiliary and consultative roles relating to the central administration:

Superior Council of Health;

Superior Council of Public Works;

Superior Council of Antiquities and Fine Arts.

There is also a Central Commission Against Atmospheric Pollution within the Ministry of Health, instituted by Law 615 of 1966 (see Chapter 2 on Air); this Commission includes representatives from all interested Ministries.

INTRODUCTION

The law on the protection of water from pollution (Law 319 of 10 May 1976) has provided (article 3) for the State functions to be carried out by a Committee of Ministers (now known as the Interministerial Committee Law 650 of 1979), consisting of the Ministers for Public Works, the Mercantile Marine and Health, presided over by the former, with the possible inclusion of other interested ministers. The function of 'technical scientific organ' to the Committee is conferred on the Superior Council of Public Works (although the Superior Councils of Health and the Mercantile Marine retain responsibilities in their own areas). The Committee also collaborates with the Superior Institute of Health and the Institute of Water Research of the National Research Council (public bodies which do not form part of the ministerial structure).

The National Research Council, besides operating through the Institute for Water Research and Studies (IRSA) already mentioned, is also involved in other sectors of research regarding atmospheric pollution, and has institutional control (as well as fairly incisive powers of decision) over scientific and technical matters including nuclear energy policy (the establishment of nuclear power stations, the quantification of and procedures for energy choices orientated towards nuclear energy etc.).

Other research and study organisations are:

ISPE (Institute for Studies of Economic Planning);

Hydrographic Institute of the Military Marine (Naval Technical Organ);

Aeronautical Meteorological Service;

LIA (Laboratory for Atmospheric Pollution).

In addition to the State organs, offices and technical bodies mentioned above, auxiliary public bodies made up of non-territorial autonomous structures of national importance also operate. ENPI (National Organisation for the Prevention of Accidents) and INAIL (National Institute for Insurance Against Accidents at Work) are relevant in matters concerning the safety of the work environment including control of machinery; ANCC (National Association for the Control of Combustion) has surveillance duties over apparatus and fuels used and participates in the Central Commission Against Atmospheric Pollution referred to above and the Regional Committee set up under Law 615 of 1966. Through its departments (equipped with laboratories and apparatus) it is able to observe violations of the anti-pollution laws throughout the national territory and to notify the Provincial Medical Officer of Health accordingly. ENEA (National Committee for Research and Development of Nuclear Energy and Alternative Energies)[1] is an organisation which operates alongside the Ministry for Industry in the control of production activities and the use of nuclear energy, and it also has control powers regarding environmental radioactivity. The Shipping Register (RINA) and the Aeronautical Register (RAI) monitor compliance with regulations regarding the operating conditions of ships and aircraft. Furthermore, some National Parks (Abruzzo and Gran Paradiso) are managed by independent bodies, while at local level there are the Regional Forestry Agencies. ENPA (National Organisation for the Protection of Animals) is a charitable body with powers of protection which began as an organisation for the protection of domestic animals but is now also concerned with the effects of pollution on wildlife.

Other authorities with powers of intervention are:

the fire service (pollution from thermal plants);

forestry keepers, at present part of the regional structure;

Customs and Excise (marine pollution);

police (public security and Carabinieri) for verification and suppression of pollution crimes.

Certain parliamentary organisations should also be mentioned which have performed an encouraging and supportive role in the past with regard to environmental protection.

(i) Special Commission for Ecological Problems (under the Senate) (*Instituted under a motion of the VI and VII Legislature*): this Commission supplied opinions on environmental protection laws under discussion, published studies and reports and collected contributions from the regions for the definition of legislative direction in ecological matters.

(ii) Committee for the Study of Water Problems (under the Chamber of Deputies): it has engaged, with the aid of technical consultants, in defining the legislative problems relating to water pollution and the conservation of water resources in Italy when the text of Law 319 of 1976 was elaborated, being also empowered to consult regional delegates.

Finally, the work of GET (Territorial Ecological Group) of the Computer Centre of the Court of Cassation should be mentioned: this group,

---

[1] Law 84 of 5 March 1982, entitled 'Modifications and integrations to Law 1240 of 15 December 1971 concerning the re-structuring of the National Committee for Nuclear Energy', in effect re-named this National Committee (the old CNEA) to ENEA, bringing changes in the law which would permit its developing alternative sources of energy (see Chapter 7 on Nuclear Energy).

supported by and consisting of magistrates, carries out valuable work of documentation by selecting sentences on environmental protection matters to be included in the data bank of the Court of Cassation.

1.3.1.2 ORGANS AND OFFICES FORMING PART OF THE PERIPHERAL STATE ADMINISTRATION

The list of State offices with powers of intervention or control over pollution is supplemented by the peripheral administration of the State.

At this level, the inconvenience already encountered in central administration is found once again; there is a lack of instruments to coordinate the activities of the various public authorities responsible for the different sections involved. This is made worse at local level by two circumstances: the impossibility of resorting to practices and institutions which perform some kind of coordinating function at the head of the administration (Research Commissions, Interministerial Committees, etc.), and the competition between regional powers and local bodies with resulting problems of coordination, not only within the State, but also between the State and the Regions.

A region operates not only as an autonomous territorial body (with its own powers and responsibilities) but also as a body of autarchic decentralisation, i.e. an instrument for the realisation of functions which, while being functions of the State, are delegated to the regions so that they can be put into effect through the use of the regional organisation.

Decree 616 of 1977 transferred the State functions relating to the Regional Committees Against Atmospheric Pollution to the regions, with a resultant change in the composition of the Committees (now made up of the State and local officials and representatives of interested bodies) and in the functions (until now to provide opinions on the location of installations, on the zoning of communes as provided for under Law 615 of 1966, on the integration of communal health and hygiene regulations, to which should be added matters of acoustical, water and land hygiene).

Law 615 of 1966 also provided for the setting up of the Provincial Commission for the Control of Atmospheric Pollution. The Committee takes advantage of its work. This is a predominantly technical organisation with functions of control and investigation of industrial plants.

Within the provincial administration the official traditionally central to Government activity has been the prefect, who is directly responsible to the Ministry of the Interior and is truly the correct channel for Government activity in this area. In addition to his supervisory powers in terms of the administrative activity of the State in the provincial area, the prefect has considerable powers, especially through certain effective discretionary provisions such as the power to issue emergency ordinances for reasons of public order and safety (which are not subject to restrictions and can make exceptions to existing law but not to the Constitution, nor the principles of the legislature, etc). The prefect can also issue contingency and emergency provisions on specific matters (e.g. building, local police, hygiene) and for reasons of public safety or health (in this case in the opinion of the Provincial Medical Officer). In matters of public safety and hygiene the prefect can substitute his provisions for those that the mayor (who has regulatory powers in these sectors) has failed to issue in his capacity as a Government official. The non-observance of ordinances issued by the prefect is punishable under penal law (administrative offences under the communal and provincial TU, article 20).

The laws on air pollution (Law 615 of 1966) and on the use of radioactive substances (DPR 185 of 1964) have conferred important powers on the prefect: he can decide on appeals against provisions which are the responsibility of the Provincial Fire Service or the mayor (civil thermal plants), he can order the closure of industrial establishments which do not comply with Law 615, he can grant *nihil obstat* (approval) for the storage of radioactive substances and the use of radioisotopes in scientific research or production activities (DPR 185), etc.

The protection of the marine environment from pollution, which at the central level is the responsibility of the Ministry of Mercantile Marine, at the peripheral level is conferred on the following bodies of the Ministry:

> Maritime authorities: administrative zones of the maritime directors, who issue provisions regarding harbour regulations and permissions relating to State-owned waters, coordinate the zonal services and have investigative powers over maritime accidents.
>
> Harbour offices: in charge of a maritime department with powers regarding the policing and safety of ports.

Inland waters, however, in addition to being the responsibility of the region, are also under the authority of some peripheral bodies under the Ministry of Public Works.

In fact, some offices of this Ministry have specific responsibilities for matters regarding public waters (for example, the river magistrates—Po, Tiber, Adige, Mincio etc.—and the magistrate for the Venetian Lagoon, with powers over the water regulations of the river basins; a further example is the Hydrographic Service with responsibilities of surveillance and observation). Other offices of the Ministry have general responsi-

INTRODUCTION

bility (i.e. they exercise ministerial powers within a given territory). The General Supervisors of Public Works and the offices of the Civil Engineers should be considered here but both these are State bodies acting as both State and regional bodies simultaneously only in so far as powers not transferred to the regions are concerned. As State bodies they are responsible for water works, State building construction, maritime and watercourse work. The Civil Engineers make recommendations regarding permits relating to State-owned waters, carry out the formalities for issuing permits, and also have important delegated powers regarding the use of public waters.

The Heads of the Provincial Fire Services are answerable to the Ministry of the Interior and are responsible for matters concerning civil thermal installations (approval of plans, testing, controls of operation, fuels and emissions). Under Law 615 of 1966 on air pollution, the Heads of the Provincial Fire Services have technical auxiliary powers for verification of the levels of industrial pollution in the atmosphere.

The peripheral bodies of the Ministry of Education are mainly important in the defence of the environment as a cultural asset in a broad sense, and with priority for certain environments which are more particularly defined. In this perspective, the Superintendents of Monuments participate in the urban and territorial organisation of a zone with their authoritative powers. Furthermore, in the context of the protection of areas of natural beauty, under Law 1497 of 1939 the Ministry of Public Information makes use of Provincial Commissions for Natural Beauty to assess the cultural/environmental assets which should be protected.

In the field of air pollution particular importance lies in the regulations concerning motor vehicles. This is the responsibility of the Ministry of Transport and Civil Aviation through the Department of Civil Motoring. At local level this is represented by the Provincial Civil Motoring Inspectorates with powers of control over motors (type approval, compulsory inspections) and conformity with anti-pollution regulations regarding safety criteria and apparatus. Air traffic is, however, controlled by offices of the Department of Air Traffic and Airports which is headed by an Office of General Direction for Civil Aviation (within the Ministry of Transport).

Regarding the quality of work environments and the safety of manufacturing processes (responsibility of the Ministry of Labour), Labour Inspectorates operate at the local level with great investigative powers, ensuring that the provisions in force are complied with. In this respect it should be remembered that mining and extraction come under the supervision of the Ministry of Industry which operates through the Mining Corps.

The Chamber of Commerce, Industry and Crafts, on the other hand, carries out a very important public function. It is a public body representing the interests of the business section of a province, but presided over by a president nominated by the Government (as such, he takes part in the collegiate bodies set up at regional and provincial level by the law against atmospheric pollution—Law 615 of 1966).

But direct participation in the public bodies of local importance for the protection of the environment is also by means of the obligatory and voluntary public consortia, which constitute bodies placed in charge of the control of certain activities within a territorial sphere. There are consortia for the protection of fishing (engaged in the surveillance and prevention of water pollution); for land reclamation (concerned primarily through their power to control the use of water); and for industrial centres.

1.3.1.3 ORGANISATIONS AND OFFICES WHICH ARE RESPONSIBLE TO THE REGIONS AND OTHER LOCAL AUTHORITIES

Following the creation of the regions with ordinary status and the decrees transferring powers in 1972 and 1977, the framework of organs, offices and bodies related to local authorities has greatly altered: some State offices have been regionalised, others are currently in a position of joint dependence on State and region, some autonomous bodies have been included in the regional sphere, and the provinces and communes have also become instruments of regional administrative activity; they are also autonomous territorial bodies and decentralised offices for certain State functions.

The following regional organs are significant because of their powers in public health matters:

The Provincial Medical Officers: these have powers to issue ordinances in cases of necessity or emergency, to punish infringements of health provisions, to close installations and confiscate products, to replace mayors who are not carrying out their duties in issuing health orders, to decide hierarchical appeals against such ordinances of the mayor, and to watch over hygiene and public health matters by issuing directives to health officials. They can make use of technical commissions on toxic gases and ionising radiations. DPR 616 of 1977 transferred some powers previously delegated by the State to the regions.

The Provincial Health Council (with consultative, technical and advisory proposal functions): this is obliged to give opinions on communal hygiene and health regulations. This body is replaced in some regions by a Regional Health Council (see Law 30 Reg. Abruzzo of 25 July 1973) which has absorbed the various provincial councils.

The Health Officials: these are regional officials hierarchically subordinate also to the commune, for which they perform a consultative function. They have responsibility for providing opinions on authorisations for the operation of unhealthy factories and for contingency and emergency ordinances relating to health matters.

The Provincial Veterinary Surgeon: he is responsible for hygiene, preventive treatment of animals and control of livestock breeding.

The law on health reform (Law 833 of 23 December 1978) which came into force on 1 January 1980 endowed the regions with important powers. It provides for the constitution of multizonal defences and services for the control and protection of environmental quality (article 15 of Law 319 of 1976 in the new text introduced by Law 650 of 1979 provides for technical functions of surveillance and control over all discharges to be handed to them).

The offices of the Civil Engineers and the Regional Superintendents for Public Works have only come under the authority of the regions with regard to transferred responsibilities (cf. 1.3.1.2).

Law 319 of 1976 on water pollution ('Merli Law') had provided among the duties of the regions (article 8) for 'the reorganisation of the peripheral technical and administrative structures in charge of the public water and sewerage services'. The plan for these should have been sent to the Committee of Ministers by every region, within 3 years of the law coming into force; it also specified that the objectives should be achieved within 10 years of that date. A new law (Law 650 of 1979, called Law 'Merli bis' extended this period by giving the go-ahead to a new phase of administrative reorganisation on the subject.

We find ourselves, however, in the presence of a phase of reorganisation of regional offices as a result of the new transfer of functions to the regions by DPR 616 of 1977 (this time by 'subject sector'). Article 105 of this DPR states that

> until the regions and local bodies have instituted their own specifically competent organs or technical offices, they will make use of the central and peripheral State organs and technical offices in carrying out the functions transferred to them in matters of protection against pollution. In order to carry out the functions delegated to them in this matter, the regions and local bodies must make use of the State organs and technical offices. It could therefore be considered that the present balance, characterised by the co-existence

of a functional dependence by the regions on certain State organisations, is destined to be altered rapidly as far as the sections of regional responsibility are concerned.

With regard to environmental protection, the reorganisation of the offices on a regional basis came about with ease, through the transfer of offices previously belonging to the State, and on account of the fairly remote origin of certain services or powers of the administration, in comparison with the relatively recent pollution sector which has still to be set up under stable organisations.

Thus, in 1972 (DPR 11 of 15 January 1972) the Forestry Inspectorate (at one time a peripheral office of the Ministry of Agriculture and Forestry) was transferred to the regions. But more generally, in this administrative sector also in connection with the regionalisation of forest lands, the State Agency for Forestry has been replaced by Regional Forestry Agencies.

These are examples of regional public agencies, i.e. organs with functional and administrative autonomy within the framework of the Assessorate responsible. But the region can also organise bodies of a legal character and act as devices instrumental to the region's own purposes. Out of all these bodies under regional authority, or 'para-regional' bodies, the 'development bodies' (with responsibilities for planning, also by means of incentives, for research into manufacturing problems, for the promotion of desirable types of production activity etc.) are of particular importance for our purposes. Some of these organs arose from the ashes of previous bodies of local importance (e.g. the Fucino Body in Abruzzo is today the Abruzzo Development Body, ESA).

The territorial bodies which operate within the region (communes and provinces) have their own responsibilities—as regional offices for those matters of regional responsibility which are administered through minor local bodies, and as Government offices for other functions.

DPR 616 of 1977 has transferred to the regions (as from 1 January 1978) the functions carried out by the mayor as Government official within the meaning of articles 216 and 217 of the TU health laws (protection of public health with reference to unhealthy factories) and article 226 (protection of food use and domestic use of waters from lakes, canals or watercourses in relation to authorisation for the opening of housing, factories, hospitals, sanatoria etc. discharging polluting waters or effluent).

The mayor maintains his function of Government official in the issuing of contingency and emergency provisions (in questions of building, local police, hygiene or for reasons of health or public safety).

INTRODUCTION

The province is important because of the number of relevant organs and technical offices operating within its territorial sphere:

> The Laboratory of Hygiene and Preventive Medicine is a technical organisation of the province for analyses, research and surveillance. Article 15 of Law 319 of 1976 (on water pollution) bestows on it 'the technical surveillance and control functions over all discharges' until the multizonal garrisons and services for control and protection of environmental hygiene are put into effect.
>
> The Service for the Surveillance, Monitoring and Control of Atmospheric Pollution is an office set up under article 7 of Law 615 of 1966 (on air pollution).

With the passage of certain decentralised State functions to the provinces in the sphere of regional responsibility, the role of the province is changing: there is a tendency to concentrate, at the regional level, some delegated powers which were previously carried out by the provinces as decentralised State administrative offices.

In addition, some important powers have been eroded by the creation (Law 1102 of 31 December 1971) of the mountain communities which are public legal bodies formed by the communes in zones delineated by regional laws responsible for the creation of economic and urban development plans.

The consortia of communes or provinces are also very important for the management of services of works of common interest. For example, Law 319 of 1976 on water pollution confers on consortia of communes (or mountain communities), as well as to single communes, the management of water services for environmental hygiene (article 6). In addition, DPR 616 of 1977 provides the regions with the possibility of creating 'common offices or managements, even in the form of consortia' for services or works involving neighbouring territories.

1.3.2 Groups with collective and widespread interests and their importance for the environment

In terms of environmental protection, bodies and organisations are considered under three main categories:

(i) Pressure groups influencing public opinion and indirectly promoting laws, administrative acts, judicial initiatives etc.

(ii) Protagonists in administrative proceedings opposing public bodies or as components represented in organs or offices with public functions, or in consultative organs.

(iii) Holders of general or widespread interests to act in civil, penal and administrative proceedings.

Regarding (iii) above, there is the problem of the protection of those individual situations which do not involve a specific owner or uniform collective interest. The problems which occur concern primarily the qualification of such widespread or collective interests: they can consist of situations which can already individually be protected, but which are common to several people, or more often, of interests which do not amount at individual level to protected interests, remaining simple interests, but assume a particular importance because of their collective dimension (e.g. the collective interest of health, which is distinct from, and requires protection in a different manner from, the individual health interest—see article 32, Constitution).

Another related problem is that of the qualification of the group representing such interests: i.e. of the conditions that constitute real representation with regard to the interests in question.

Both of these problems combine to constitute the problem of the ability to protect widespread or collective interests in the face of article 100 of the Code of Civil Procedure which restricts recourse to legal action to the person or persons to whom the individual interest applies (the bodies in question would not hold an interest of their own, nor necessarily of their members). Such a problem, to which the decision of the courts has already given reassuring answers, will appear clearer when it is reinforced by the treatment of situations which can be protected and in the procedural profile (see 1.4).

Groups coming under (ii) above are of considerable legal interest because of the opportunities at their disposal for participation in the formation of public decisions.

Consultation with groups and associations on the part of public authorities is a very common practice (e.g. the participation of the trade unions in administrative organisations, invitations for representatives of Confindustria (Confederation of Industry) on interministerial commissions or the central commissions for atmospheric pollution, and the provision for consultation with representatives of manufacturing associations, to give evidence to parliamentary commissions—article 3 of Law 615 of 1966. It is also a practice which receives legislative acceptance (Law 63 Prov. Bolzano of 1973 on water protection: mandatory consultation with the League of Farmers of the Alto Adige region).

INTRODUCTION

Groups of importance in this sector can be defined according to their differing attitudes to the problem of pollution:

(1) Groups which are only incidentally concerned with environmental protection (trade unions, Confindustria, etc.). They are nevertheless of considerable importance both because of their size and organisation and because they perform important duties: the 'Workers Statute', for example, bestows on the unions duties of healthiness control and research concerning the work environment and the healthiness of the manufacturing processes; this has led the union also to take an interest in pollution produced in the external environment of the factory, but this is still only a sporadic occurrence.

(2) Groups interested in activities or sectors relating to the natural environment, such as CAI (Italian Alpine Club), ATI (Italian Thermo-technical Association, which also has a representative on the Central Commission for Atmospheric Pollution), and the various Associations of Farmers.

(3) Naturalist Associations: These are bodies which have evolved for the protection of the environment as a cultural asset of the countryside: for example, Italia Nostra, a recognised association founded in 1955 and often at the centre of legal battles in which its legal powers to act in the broad protective interest have been discussed and at times recognised. This grouping also includes ecological associations of national and international interest (e.g. the Italian section of the World Wildlife Fund) and groups of minor importance (such as the Federation for Nature etc.), down to groups of a predominantly political nature similar to the ecology parties in other countries (e.g. Kronos 1991), but otherwise of little significance at present.

1.4 PROTECTION OF THE RIGHTS AND INTERESTS OF INDIVIDUALS

1.4.1 Interests of importance in terms of environmental protection

Here we must describe the means of protection available to individuals regarding environmental protection. In this respect it is necessary to define the legal definition of a private interest. There are two subjective situations which are considered to be legally actionable: the subjective or individual right and lawful interest (article 24, Constitution: 'Everyone has the right to go to law to protect their own rights and lawful interests').

The individual right is included in many legal systems and concerns an individual interest which is specifically protected, so that any injury to such an interest in a direct and immediate manner falls within the legal sphere of the owner of the right in question.

In cases of lawful interest, which is a peculiarity of Italian legislation, the interest of the individual is protected provided it reflects the general interest in the correct functioning of administrative activity, and, in short, observance of the law. Interests which are occasionally protected, or reflect other interests, have been discussed: the important point is that in the face of the lawful interest (unlike individual right) there is no obligation or positive or negative duty, but an active situation, i.e. (discretionary) administrative powers. The misuse of these powers or the exceeding of imposed limits harms general interests and not the individual interest. However, where an individual bears an interest which does not coincide with the general interest, but is particularly qualified for protection (and not 'simple') he could, as the bearer of a 'lawful interest', appeal, even judicially, against the harmful act.

The importance of the distinction also rests in the fact that it confirms the jurisdictions of the ordinary law judge or the administrative judge: in Italy a dual system is in force by which protection against the public administration is the responsibility of different judges depending on whether they act *iure imperii* (in an administrative capacity) or *iure privatorum* (a private capacity). The ordinary law judge is qualified to recognise harm to individual right even if committed by the Public Administration. Harm to lawful interests, however, is reserved for the jurisdiction of the administrative judges (see 1.4.2).

However, not every interest can constitute an individual right or a lawful interest: in the first case expectations of specific protection are required for the interest of an individual (who at the same time is the bearer of the interest and the subject of protection); in the second case the position of the individual must be such that even though he is not the bearer of the general interest in the proper functioning of the administrative activity, he can nevertheless complain of injury to a different, but particularly eligible, individual interest. There are therefore some (simple) individual interests that cannot be protected since they are not sufficiently legally defined but are nevertheless important because of their collective or general dimension. This is the case with environmental protection matters, a privileged area emerging from the theme of widespread interests. These are presented, as we have seen (1.3.2) not only, or not as much, in the case in which the same protected interest (even at the same level of individual right) is common to a large number of people, but also and above all when individuals have only a slight interest which nevertheless reflects, at the general and collective level, an interest of great importance meriting ample protection. Here the importance of legalising the activities of associations, groups and bodies representing such interests (which would otherwise remain judicially unrepresented) is made clear. The decisions of the court appear to be oriented towards the recognition for associations which pursue aims of environmental protection and of their right to act as representatives of general rather than individual interests (see the judgement of Council of State, 9 March 1973, b. 253, in *Riv. giur.* ed 1973, I, 2, 261 regarding the 'Italia Nostra' association); recently there has been a judgement passed by the Trib. Salerno, 6 February 1979, which in principle recognised the possibility of Italia Nostra and the WWF acting as a plaintiff in penal trials; in legal writing the problem has been added to with reference to the connected themes of the protection of the consumer (RUFFOLO) and the protection of the environment (CORASANITI).

1.4.2 Legal bodies

The offices are organised as follows:

(i) Civil jurisdiction
Conciliators (first instance judges with limited competence in terms of value and subject matter)
Magistrates (first instance judges for limited conflicts and for appeals from judgements of the Conciliators)
Tribunals (first instance judges for controversies not covered by the Magistrates and Conciliators and for appeals from the Magistrates)
Court of Appeal (appeals from the Tribunal)
Court of Cassation (third instance judges handling only 'points of law')

(ii) Penal jurisdiction
Magistrates (judges with inferior competence)
Tribunals (judges with intermediary competence and also competence for appeals from Magistrates)
Court of Assize (for more serious crimes)
Appeal Court of Assize (only appeals from the Courts of Assize and Tribunals)
Court of Cassation (third instance judges handling only 'points of law')

(iii) Administrative jurisdiction
 Regional Administrative Tribunal (TAR) (first instance judges competent for acts effective in the regional sphere; the TAR of Lazio also handles acts relating to the central administration)
 Council of State (judges of second instance)
 Territorial Tribunals for Water and Superior Tribunals for Public Waters (competent for public water permits)

The Court of Cassation is the supreme penal and civil court responsible for watching over the formal and substantial correctness of judgements passed by the various judges: it is also the court of jurisdiction, i.e. it settles questions relating to the type of judge (civil or administrative) who should decide the case in question.

In protecting an individual right, the civil judge can pass a judgement of verification (which affirms the existence of a certain legal relationship), a constitutive judgement (which alters, creates or terminates a legal situation), or a judgement of condemnation (which imposes performance). The latter may be followed by an executory process in the case of non-compliance with the sentence. He cannot annul ('revoke or modify') an administrative act (article 4(E) of Law 2248 of 1865) but he can ensure that the judgement is not applied in a particular case if he feels it to be unlawful (article 5 of the above law).

The administrative judge, on the other hand, can annul or amend an unlawful administrative act but cannot sentence (e.g. in cases of compensation for damage the only aspect over which he has this kind of jurisdiction is the question of legal costs). This system is of doubtful efficiency since it makes further civil action necessary.

In the penal process the action is tried by the Public Prosecutor which is a State body belonging to the judicial organisation. The Public Prosecutor is required to exercise penal action on the basis of notification of a crime, regardless of how it has been received. However, for offences which can only be prosecuted following a complaint, this complaint must be lodged for such an action to be prosecuted. The penal judge can condemn the offender to a term of imprisonment, or a fine for an offence and detention, or a fine for an infringement. In other cases the defendant can be acquitted if the act did not take place, if it was not committed by the accused, if there is insufficient evidence, or if the act does not constitute an offence or is not provided for under the law as an offence.

1.4.3 Authority, powers and rights of the individual (to information, to oppose authorisations or permits, to indictment, to obtain the cessation of harmful activities or to compensation)

When protecting their rights and interests individuals can therefore apply to the legal authorities. Whenever an individual complains of damage he can apply to the civil authorities for compensation, and claim also against the Public Administration where damage is caused by material or legal activities on the part of public functionaries or their representatives. In these cases (article 28, Constitution) there is joint responsibility of the administration and the perpetrator of the act (but the injured party cannot act against the latter in a case of a slight offence only) providing there is a connection between the act committed by the functionary or representative and the activity of the administrative body concerned.

Compensation for damage can also be applied for to the penal authorities by acting as plaintiff. According to prevailing doctrine, the injury must be direct and to a private individual; social damage does not qualify in this context. Thus the possibility of protecting collective and widespread interests represented by groups and associations by acting as plaintiff in penal proceedings is limited, even though there are examples which contradict this premise (P. Viareggio 10 October 1972). A notable breakthrough is constituted by the possibility, asserted by the Court of Cassation, for civil and public bodies to act as plaintiff to gain compensation for 'social, ecological and biological damages leading to a deterioration of the urban environment and a decrease in the economic value of the communal territory' (Cass. 5 July 1974). On the subject of penal proceedings, another power given to individuals and often successfully used by naturalist and ecological associations is the power of notifying the public prosecution of a crime by an indictment; in addition, the victim of a crime involving pollution has the right to bring an action or apply to the Public Prosecutor in the case of offences which require such action before prosecution can be made. We should remember that once the Public Prosecutor has been informed of a criminal activity (*notitia criminis*), penal action is obligatory.

In administrative proceedings the individual can have harmful administrative action annulled or amended (see 1.4.2). However, in principle, such action remains effective on presentation of the request for annulment or amendment, and in fact preserves all its effects until the outcome of the proceedings is made known. It is, however, possible to propose a preliminary petition to suspend effectiveness of the action under special circumstances or conditions.

In addition to legal remedies, there are other means for the rejection of administrative actions operating through administrative channels for the protection of individuals against harmful provisions of the Public Administration. It is possible to make use of a 'hierarchical appeal' to a higher authority than that which ordered the administrative action in order to obtain the annulment or amendment of the action in question. The lodging of an appeal is at times considered necessary so that, if the appeal does not have the desired outcome, it will later be possible to resort to jurisdictional action. In exceptional circumstances an appeal can be made to the President of the Republic but this is controlled by special procedures and only admitted in specific cases.

However, the most interesting aspect of the protection of individual interests in the realm of administrative activity is not the means at the disposal of individuals once the administrative action has been undertaken, but the formation of the action itself: the ever greater bureaucracy within the administration has led to participation by the greatest possible number of people in the formation process of administrative actions.

The inadequate 'transparency' of administrative activity is made clear above all with regard to the instruments of publicity for administrative actions: today they are not so much aimed at the dissemination of information to the public but rather at guaranteeing the correctness of certain relationships with regard to individuals (public announcements of tenders, public auctions etc.). However, there is one aspect of administrative activity of importance for our purposes in this context, and that is the publication of Regulatory Plans, plans for industrial zones and territorial countryside plans etc. Such publications are undoubtedly mainly a formality but can represent a means of satisfying the private interest in the right to information.

In the sector of town planning instruments there is also recorded the only example of the right of observation and notification attributed to individuals in regard to Public Administration decrees and the Public Administration obligation to take such objections into consideration to a certain extent (normally a notification to the administrative authorities does not oblige them to respond).

In fact, not even in the sole case mentioned above (articles 9 and 15 of Law 1150 of 1942) is there an absolute obligation on the part of the Public Administration to respond, although if one such observation is under examination, the Public Administration must take into account all observations on the matter in the subsequent deductions.

2
Air

2.1 GENERAL PROBLEMS

2.1.1 Introduction

The controls relating to air pollution consist of relatively recent legal provisions; for example, Law 615 of 13 July 1966 and its implementing regulations DPR 1391 of 1970, 323 of 1971 and 322 of 1971, for the following sectors: thermal installations, diesel powered vehicles and industrial installations; and Law 437 of 3 June 1971 on exhaust pollution from motor vehicles with positive ignition. There are also provisions which are somewhat distant in origin but nevertheless important because of their far-reaching powers which are accorded to the public authorities (see articles 216 and 217 of the TU of the health laws and articles 152 and 153 of the RD of 4 February 1915, no. 148 TU of the communal and provincial laws, and article 20 of RD 383 of 3 March 1934). The provisions under the Civil Code and the Penal Code are also significant (especially article 844 of the Civil Code and article 674 of the Penal Code); these have only been partially outdated by the discipline which has come with Law 615 of 1966 (already mentioned).

The relationship between the 'new' discipline produced on the basis of Law 615 of 1966 and that previously in use raises two distinct questions: that of the real value of the innovation introduced by Law 615, and the fate of the methods of protection based on the regulations contained in the Civil Code or the Penal Code in use before the law was introduced (see 1.2.3.1).

As regards the former, it should be emphasised that from the law in question can be deduced a complete legislative definition of air pollution can be deduced: damaging alteration to the components of the atmosphere as a result of emissions of smoke, gas, odours, dust or fumes. It also states that those harmful changes which cannot be avoided must be contained. The law is thus seen as attempting to achieve the protection of clean air as a right in itself (ROBECCHI MAJNARDI). Also to be pointed out is the modernity of the legal system in limiting itself to laying down the basic controls while leaving the details to the implementing regulations (of which, however, only those applying to thermal installations appear to be sufficiently comprehensive; the industrial installation and vehicle sectors have attracted the most criticism—see AMENDOLA on this last sector).

As far as the latter of the two questions is concerned (i.e. the relationship with previous civil and penal provisions), the problem is posed with regard to the possibility of applying article 674 of the Penal Code (which punishes emissions of gas, vapour and smoke apt to offend, soil or molest persons) to those areas covered by Law 615 and also where the polluting activity violates the law itself (and therefore comes within the scope of article 20 of that law, which makes the omission of anti-smog methods or precautions a punishable offence).

The decision of the courts has upheld the applicability of article 674 of the Penal Code, both because this article and article 20 of Law 615 protect different legal rights (the former protects the individual and his integrity and the latter the purity of the atmosphere—see Cass. 20 December 1972, in *Giust. pen.* 1973, II, 54) and because they apply to different types of action (article 674 punishes those who emit offensive gases, fumes etc., whereas article 20 of Law 615 punishes those who fail to take the anti-smog precautions required by the law). See Cass. 20 December 1972, in *Giust. pen.* 1973, II, 54; Cass. 31 October 1975, in *Mass. pen.* 1977, 1371; Cass. 1 February 1977, in *Giust. pen.* 1977, II, 414; Cass. 10 January 1977, in *Giust. pen.* 1977, 628; T. Torino, 18 March 1976, in *Giur. it.* 1976, II, 628; P. Dronero, 30 May 1973 in *Guir. it.* 1976, II, 223; P. Trento, 11 December 1975, in *Foro it.* 1976, II, 194.

There is therefore a tendency to consider the emission limits provided for by the law as relating to the effects of the law itself and not bearing on the consideration of whether or not these emissions are tolerable or constitute a punishable offence. P. Ferrara, 31 January 1975, in *Giur. it.* 1976, II, 172, considers article 844 of the Civil Code to be violated even when the limits under Law 615 have not been demonstrably exceeded. Rather, it considers, in cases where mobile or stationary objects are rendered unusable, that this is a punishable offence under article 635 of the Penal Code. This last regulation is in addition to—and not as an alternative to—those introduced by the above-mentioned law.

Another important question is that of the relationship between the anti-smog legislation and the requirements of the public health laws, which are far older. It has on the one hand been considered that Law 615, which seeks to control atmospheric pollution by a global approach to the problem, has removed this subject from the sphere of activity covered by article 216 of the TU of the public health laws (which provides the mayor with ample discretionary powers to assess the adequacy of technical equipment installed to prevent injury to health, as well as the ability to order the installation of plant in residential areas).

We should therefore now pass on to an assessment of the following of hygiene and health requirements only as far as atmospheric pollution is concerned, in relation to the criteria contained in Law 615 of 1966, which preserves only to a limited extent the mechanism of article 216 of the TU of the public health laws with regard to protection other than from air pollution (as pronounced by the State Council in consultative session on 17 October 1972, n. 582).

The decisions of the court state that the provisions in existence prior to Law 615 of 1966 are still in force with regard to the protection of all aspects of hygiene and health, excluding atmospheric protection; the responsibilities of the mayor under article 216 TU of the public health laws therefore remain in force (T.A.R. Liguria, 4 May 1978, no. 207, in *Trib. amm. reg.* 1978, I, 2744; T.A.R. Lazio, section II, 10 March 1976, no. 177, in *Trib. amm. reg.* 1976, I, 1188).

With reference to this interpretation of the provisions in question, it has been noted that while from the point of view of legal logic the separation already described between the two aspects of protection appears unexceptional (Law 615, a specialised law, removes the atmospheric pollution aspect from the responsibility of the general regulations—article 216 TU of the public health laws), nevertheless if those general remedies granted under the public health laws were no longer available to individual members of the public, it could lead to a decrease in overall protection, in cases of danger or of damage to public health not covered by the specific provisions of the 'anti-smog' law. But in this respect there are other judicial pronouncements which recognise the power of the mayor to adopt contingency and emergency measures under article 153 TU 148 of 1915 (Communal and Provincial Law) in the event of dangerous situations. This is also covered by article 4, no. 8, DPR 322 of 1971 (implementing regulation of Law 615 of 1966 relating to the industrial sector) which preserves the powers of the mayor in article 153 TU already mentioned (T.A.R. Toscana, 13 November 1975, no. 397, in *Trib. amm. reg.* 1976, I, 196). This last point will be discussed further in 2.2.6.

2.1.2 Control by land-use planning

There are two distinct methods of land-use planning; which method applies depends on whether the activity concerned is a human activity, more or less productive, but which in any case is liable to cause incidents which will affect the environmental balance or is directly related to the problem of air pollution.

2.1.2.1 GENERAL

See 1.2.2.2 and 1.2.2.3 regarding instruments of urban planning and economic programming.

The restrictions relating to landscape planning, hydrogeology and land conservation also contribute indirectly to the protection of the atmosphere. Law 1497 of 29 June 1939 was brought in to protect certain areas identified by provincial commissions (1.3.1.2) and designated areas of natural beauty, by provisions of the Ministry of Education, from any changes or modifications which alter the natural characteristics of the areas. The regions with ordinary statutory powers are able to formulate a territorial landscape plan which defines the set of rules under which particular zones are to be placed. In spite of the fact that the approval mechanism for this plan, formerly the responsibility of the Ministry of Education, is now identical to that in use for urban planning (by decree from the President of the Republic); the protection of areas of natural beauty are still held to be under two distinct spheres of responsibility; the former on a regional and the latter on a national level (see Corte Const. 142, 1972).

The hydrogeological restrictions contained in Royal Decree Law 3627 of 30 December 1923 limit, and in some cases prohibit, any change in land use which might result in damage to the public and in the loss of ground stability or disturbance to groundwaters. Mountainous and forested areas are therefore subject to special restrictions. The law is, however, too outdated for today's requirements and of only limited effectiveness in dealing with offences. In addition, it is not properly connected to the urban planning and land protection disciplines.

2.1.2.2 PLANNING PROCEDURES UNDER 'ANTI-SMOG' LEGISLATION

Law 615 of 1966 divides the national territory into two zones, A and B, and the controls contained in the law apply only to areas within these zones, thus creating a precise territorial distinction between the regulations under the 'anti-smog' law and those contained in the TU of the public health laws. The latter still applies to atmospheric pollution in areas which are not zoned under the various Ministry of Health decrees; see Cass. 4 May 1973.

For the purpose of preventing atmospheric pollution, the national territory is subdivided into two control 'zones' called zone A and zone B.

Zone A set up by Law 615 of 1966 includes:

(i) Communes of central and northern Italy with a population of from 70,000 to 300,000, i.e. with a smaller population but with particularly unfavourable industrial, urban, geographic or meteorological characteristics with regard to atmospheric pollution, according to the opinion of the Central Commission under article 3.

(ii) Communes of southern Italy and the islands with a population of from 300,000 to 1,000,000, i.e. with a smaller population but with particularly unfavourable industrial, urban, geographic or meteorological characteristics with regard to atmospheric pollution in the opinion of the Central Commission.

(iii) Localities which, in the opinion of the Central Commission, are of particular public interest.

Zone B includes:

(i) Communes of central and northern Italy, even those with a population exceeding 300,000 and communes of southern Italy and the islands with a population exceeding 1,000,000.

(ii) Communes as mentioned above, with smaller populations, provided they have (in the opinion of the Central Commission) particularly unfavourable industrial, urban, geographic or meteorological characteristics with regard to atmospheric pollution.

The distribution of the communes concerned into the two zones under this article is carried out by decree from the Minister of Health, on the opinion of the Central Commission. The Minister of Health can also include a commune in one of the two zones, regardless of the number of inhabitants or geographical location, on receipt of a suitably justified request.

If an area has already reached the pollution limit laid down by law and cannot tolerate an increase in industrial emissions, the commune and the province have the right to notify the Regional Committee for Atmospheric Pollution, which in its turn can make a request to the administrative authority to prohibit further industrial development (the Committee itself has no powers of direct intervention).

In the case of marked levels of high pollution, the Regional Committee can propose to the Central Committee that a commune already included in zone A be transferred to zone B. The supervisory work of the Regional Committee has often prevented, particularly in the north, new industrial installations being constructed in areas which are already congested from an environmental point of view.

The possibility of displacing industrial installations to certain areas of the communal territory, at any rate to areas outside the inhabited areas, we will discuss with regard to the rules regulating such installations.

The siting of installations producing electric power (other than thermoelectric power stations) comes under Law 880 of 18 December 1973.

CIPE (the Interministerial Committee for Economic Programming) first consults the Inter-regional Consultative Commission, assesses the suitability of the proposed location and then approves the long-term plans for the construction of new power stations. At this point it is the responsibility of the region, in consultation with ENEL (National Corporation for Electric Energy), and in agreement with the communes concerned to determine the siting of the new power station. This is done not merely on the basis of technical and economic criteria, but also on the basis of environmental protection requirements (CIPE is empowered to select the site if the region has failed to do so within the deadline imposed by the law: 3 months must elapse from the approval of the long-term programme, plus a further 2 months before CIPE can take over this responsibility from the region). The authorisation of the installation is granted by the Ministry of Industry, in consultation with the Central Commission for Atmospheric Pollution, once the previous *nihil obstat* has been obtained from the Superintendent of Antiquities and Monuments and a favourable opinion from the Ministries of Health, Environment, Education and from the President of the region. The plans for the installation cannot be approved unless they contain indications concerning measures to prevent damage to public health and to the environment.

We shall discuss the siting of nuclear power installations (Law 373 of August 1975) in the section relating to nuclear energy.

When considering in detail the anti-smog provisions it should be pointed out that the division between zones A and B under Law 615 of 1966 refers only to stationary sources of pollution (which will be dealt with in the following section) and, in addition, the thermal installation controls apply indiscriminately to the two zones, but are, however, confined to installations with a capacity exceeding 30,000 kcal per hour.

2.2 STATIONARY SOURCES OF POLLUTION

2.2.1 Industrial installations

Law 615 of 1966 deals with industrial installations in articles 20 and 21, confirming the relevant provisions under article 216 TU of the health laws. It is therefore advisable to deal first with this source of legal provisions.

Article 216 deals with factories and manufacturing processes producing vapour, gas and other emissions which are unhealthy or dangerous to health. It requires them to be listed in two classifications compiled by the Higher Council of Health and approved by the Ministry of Health in consultation with the Ministry of Industry (the same procedure is followed for revisions or modifications to the list, which are normally conducted every 3 years).

The first of the two classifications contains industries which must be sited in open country or are only permitted in a built-up area if the authorities can be satisfied that the manufacturer has demonstrated that the adoption of certain suitable technical measures has eliminated the danger to public health.

The second classification comprises manufacturing processes permitted on the urban perimeter but which require special precautions to ensure the safety of third parties.

On the basis of this arrangement the mayor has the power to authorise the siting in built-up areas of industries falling into the first classification (recognising the possible risks involved); to classify industries operating in the communal territory according to the ministerial list; to assess the suitability of the siting of industries in relation to their classification; to prohibit the start-up of an industry included in the list; or to subject it to special precautions. In this, the mayor is advised by the Health Official.

In addition to this classification of industries and consequent attributions, article 217 TU of the health laws empowers the mayor to issue provisions which will apply in cases of danger or damage to health caused by vapour, gas or other emissions of industrial origin (see 2.2.7).

This provision is in force throughout the national territory and completes the requirement in Law 615 of 1966 which defines zones A and B (see 2.2.1 on the relationship between these two provisions).

Article 20 of Law 615 requires, on the part of the enterprises, the possession of plants, installations or protective measures 'such as to contain within the strictest limits allowed by technology, the emission of smoke, gas, dust or emissions which, besides constituting a danger to public health, can also contribute to atmospheric pollution'. The concrete limits with which to compare the violation of such an order are contained in the implementing regulation of the law (DPR 322 of 1971). This regulation lays down the requirements for anti-pollution equipment: conformity with the technological principles of operation specifically adapted to the physico-chemical characteristics of the polluting agents; greater degree of efficiency and technical protection in relation to the type of industrial installations to which the equipment is attached requires the anti-pollution equipment to be planned and constructed at the same time as the industrial installation to which it is linked.

The regulation also prescribes the use of reserve anti-pollution equipment in cases of interruptions in the functioning of such equipment as a result of breakdown, poor maintenance or other technical reasons, and requires that in cases where no reserve installation is available, any interruption in the functioning of anti-pollution apparatus should result in the closure of that part of the industrial installation which is affected. The director of the works is also obliged to inform the communal authority of variations in times and frequency of maintenance operations (times and frequency are indicated in the plans for the anti-pollution installation), interruption caused by accidental malfunction, etc.

The procedure laid down for the necessary anti-pollution equipment is as follows: simultaneously with the request for a building licence (today an authorisation) for the factory, the applicant must present a technical report describing the anti-pollution equipment they intend to install, how it works, the characteristics of the industrial installation to which it is linked, the quantity and quality of emissions, etc. This report is also compulsory in the event of any significant extension or alteration to existing installations. The mayor sends the documents to the Regional Committee for their opinion. The Committee can request further information if necessary.

The Committee can give a favourable opinion, suggest alterations or give a negative opinion: in the last two cases the mayor communicates the outcome to the applicant so that he can carry out any necessary alterations (making approval conditional on this) or present a different plan. The approval document should contain the emission limits proposed by the Regional Committee and the frequency of surveys. Article 5, clause 4 of DPR 322 of 1971 establishes that all industrial installations whose anti-pollution equipment plans have been approved must also obtain permission from the mayor before starting up the installation.

As far as powers of supervision and surveillance under article 20 of Law

615 and article 65 of DPR 322 of 1971 are concerned, see 2.2.6. See this section also for the powers of the public authority and also for the question of punitive measures to be taken when the anti-smog regulations are not respected.

See 2.2.4 for the determination of limits for permitted emissions of industrial origin. Here it is sufficient to state that since authorisation to start up an installation does not exempt it from the obligation of complying with all the anti-smog requirements (article 5, DPR 322 of 1971) so the authorisation itself does not in any way diminish the responsibility of the polluter in the event of illegal emissions (see Cass. 1 February 1977, in *Giust. pen.* 1977, II, 414; for the theory that authorisation for a noxious industry to start work under article 216 TU of the health laws does not exempt it from its penal responsibility for emissions see Cass. 21 February 1977, in *Foro it.* 1978, II, 409). The respecting or otherwise of the limits set by the provisions does not even have to be proved, because the factory is responsible according to certain court decisions (P. Ferrara, 31 January 1975, in *Giur. it.* 1976, II, 172; T. Torino, 18 March 1976, in *Giur. it.* 1976, II, 628; P. Dronero, 30 May 1973, in *Giur. it.* 1976, II, 223; P. Trento, 11 December 1975, in *Foro it.* 1976, II, 194).

Finally, article 21 of Law 615 of 1966 states that requirements for protection from atmospheric pollution must be taken into account in the formulation of communal, intercommunal or interprovincial planning schemes, correctly siting industrial zones or districts in relation to residential zones.

2.2.2 Thermal installations

Among the sectors covered by Law 615 of 1966, the one relating to thermal installations has given rise to extended regulations. As already mentioned, this law applies to installations with a greater capacity than 30,000 kcal per hour, regardless of whether they are in zone A or B, provided they are not part of an industrial production cycle (or at least, not primarily for industrial use). The matter is dealt with under headings II, III and IV of Law 615 of 1966. In this section we shall only discuss heading II and in part heading IV, and the remaining aspects will be dealt with in 2.2.5 (fuels) and 2.2.6 (powers of the public authority, particularly as regards surveillance and control).

DPR 1288 of 24 October 1967 was issued as an implementing regulation of limited content (controlling fixed installations completed after the regulation itself came into force, and setting out the requirements for the adjustment of existing installations). This was followed by DPR 1391 of 22 December 1970 which lays down instructions for the entire question of thermal installations.

The regulation above all sets out the technical requirements of the installations (heading II), establishing that premises where furnaces will be located must be ventilated by means of openings connecting directly with the outside, specifying the section of these openings in relation to the size of premises and the capacity of the installation. In addition, all openings which are not for ventilation purposes must be fitted with shutters which are suitable for preventing the emission of smoke, vapour, dust or gas. The regulation also requires that premises used for fuel storage should be ventilated by suitable openings and isolated from the outside as far as smoke, gas and vapour are concerned. Tanks for the storage of liquid fuels must be in perfect condition and equipped with a hermetically sealed opening through which samples can be taken and must have loading openings and leak tubes of diameter proportional to each other. Samples must also be obtainable through the feed pipes of the burners. The requirements controlling chimney specifications are of great importance (height, diameter etc.), for which the regulation dictates the following formula (chimney with natural draught):

$$S = K \frac{P}{H}$$

where S = area of the upright section of the chimney (cm^2); P = capacity of the furnaces (kcal per hour); H = height of the chimney from the horizontal line in the middle of the flame to the outlet into the atmosphere (m); K = coefficient equivalent to 0.03 for solid fuels and 0.024 for liquid fuels.

Article 6 number 6 ff. provides for enlargements and corrections to the chimney dimensions as fixed above.

For further requirements regarding chimneys, see 2.2.3.

The requirements laid down for furnaces are contained in article 9 (volume of combustion chambers; temperatures of smoke at outlet; compulsory and easily visible indication of the maximum capacity permitted and the type of fuel to be used; visibility of flame etc.).

Similar requirements are laid down in article 10 for the burners of furnaces of installations using liquid fuel (and the relationship between the capacity of the burners and the furnace is categorised).

The maximum period for a stoppage of fuel feed to the burners in the absence of flame is also laid down:

Maximum capacity	Maximum period of stoppage (seconds)
Up to 200,000	20
From 200,000 to 600,000	10
Over 600,000	5

An application supported by documentation demonstrating conformity with the above requirements must be presented for the construction of a thermal installation. According to article 9 of Law 615 of 1966, such an application must be presented to the Chief of the Provincial Fire Service, with the right of appeal to the Prefect in the event of non-approval (the Prefect's judgement in these cases is final). The construction, alteration or extension to installations without authorisation is punishable by a fine. Similar controls (notification to the Fire Service, appeal to the prefect whose decision is final, fines for infringement) are established for testing, which must be requested within 15 days of work completion, by notification to the Fire Service and carried out by them as stated in the regulations (the applicant has 30 days to appeal against a negative decision).

The provision of article 16 of Law 615 of 1966 is of great importance since it requires all operators of thermal installations with a capacity greater than 200,000 kcal per hour to obtain an operating licence from the Provincial Works Inspectorate. This licence is only granted to operators who have passed the examination which takes place at the end of a special training course on the operation of thermal installations.

The operation of a thermal installation by a person who has not obtained the licence (and is not included in the register of qualified operators under the requirements of article 17 of Law 615 of 1966 at every Provincial Works Inspectorate) is punishable by a fine (article 18 of Law 615 of 1966).

The implementing regulation (article 2 of DPR 1391 of 1970) provides for the coordination of this control with the former provision which already recognises the qualification to operate steam generators under RD 824 of 12 May 1927, distinguishing between two categories of installations: one (first grade licence) for thermal installations, for which a certificate of proficiency in operating steam generators is required, and the other (second grade licence) for installations for which this certificate is not required. The licence can be revoked in cases of habitual infringement under article 15, clause 3 of Law 615 of 1966 (emissions of smoke with a higher polluting content than the permitted limit). The regulation requires the Fire Service (which is responsible for surveillance—see 2.2.6) to inform the Provincial Works Inspectorate of any such incident.

2.2.3 Treatment before discharge and methods of discharge

We will here deal with our subject with reference to:
(1) industrial installations;
(2) thermal installations;
(3) surveillance activities to be carried out by the discharger.

2.2.3.1 INDUSTRIAL INSTALLATIONS

In the industrial installation sector methods of discharge and pretreatment techniques are not subject to a detailed control but only to certain general directions. Article 3 of DPR 322 of 1971 requires anti-pollution equipment to be suited to the chemical and physical characteristics of the pollutants to be treated, in order to contain the polluting emissions within the narrowest limits which up-to-date technology can achieve. This constitutes a limit to the choice available to the contractor and an assessment criterion for the controlling body of the Regional Committee in approving the plans for the installation. In this respect article 4, No. 1 is complementary since it requires anti-pollution equipment to be operated in such a way as to 'guarantee that under any operating conditions of the industrial installation . . . the emission limits fixed in the document approving the plan will be respected'.

Similarly, with regard to the method of discharge, the Regional Committee can (under article 10, clause 4 of DPR 322 of 1971) request that alterations be made to the plans submitted for approval (it can, for example, increase the height of the chimneys), thus from time to time limiting the discretionary powers of the contractor. The provision under article 3, clause 4 of the DPR should also be mentioned, by which

> the anti-pollution installations operating on a wet-cycle principle resulting even partially in continuous or intermittent discharge deriving from the process adopted, are only permitted if the liquid discharge is collected and treated in a treatment plant conforming to the regulations in force.

(See Chapter 3 on water pollution.)

Recently Law 42 of 9 February 1982 was approved in Parliament which empowers the Government to issue provisions for the implementation of some EEC Directives. Consequently, EEC Directive EEC/439/75, regarding the disposal of used oils, included in the list in the draft law 1039, will also shortly be implemented.

2.2.3.2 THERMAL INSTALLATIONS

In the thermal installation sector, in addition to the simple powers of the controlling authorities, there are certain provisions which apply in special situations.

Thus, article 13 of Law 615 of 1966 states that adequate heating apparatus is obligatory for oils with a viscosity of more than 4 degrees Engler. As far as methods of discharge are concerned, the regulation sets out important criteria regarding the structure of chimneys (article 6) and of smoke flues (article 7); some of them concern the safety and functioning of the installation (construction materials, structure, resistance to damage from collision and use, openings for the collection of inspection samples, gradients of lengths of flues, cladding, joints etc.), while others concern correct maintenance (opening for cleaning, chambers for the collection and discharge of solid waste etc.), and, finally, some relate to the method of discharge into the atmosphere, principally by establishing the height of the chimney. The height of the chimney, as well as being determined by the relevant section examined under 2.2.2 (see article 6.10, DPR 1391 of 1970, for decreases in height from which value H in the formula is obtained), is determined as follows: the height of an outlet situated between 10 and 50 metres from any opening in a residential building must be no lower than the upper edge of the highest opening, except where special permission has been granted by the commune in accordance with communal hygiene regulations and with the prior opinion of the Regional Committee. But in any case the height must never be less than that indicated for outlets situated 10 metres or less away from any opening in a residential building, in which case they must be at least 1 metre higher than the roofs, parapets or other obstructions; however, the chimney cannot rise more than 5 metres above the roof. Finally, the height of chimneys is determined indirectly according to the relationship between emission limits (see 2.2.4) and the height of the chimney as shown in the illustration on page 51.

Article 8 of DPR 1391 of 1970 is concerned with devices for the treatment of smoke and prohibits the use of wet-cycle apparatus which results in even partial discharge of substances deriving from the process into public drains or water sources.

The various possible devices for the treatment of smoke must be easily accessible, must correspond to the requirements for chimneys, relating to outlets, distances, structure, materials, internal walls, and should periodically be cleaned of residual deposits. This cleaning operation should be carried out in such a way as to prevent these deposits from dispersing and so that they can be discharged in a suitable place or else delivered to the public cleansing services.

STATIONARY SOURCES OF POLLUTION

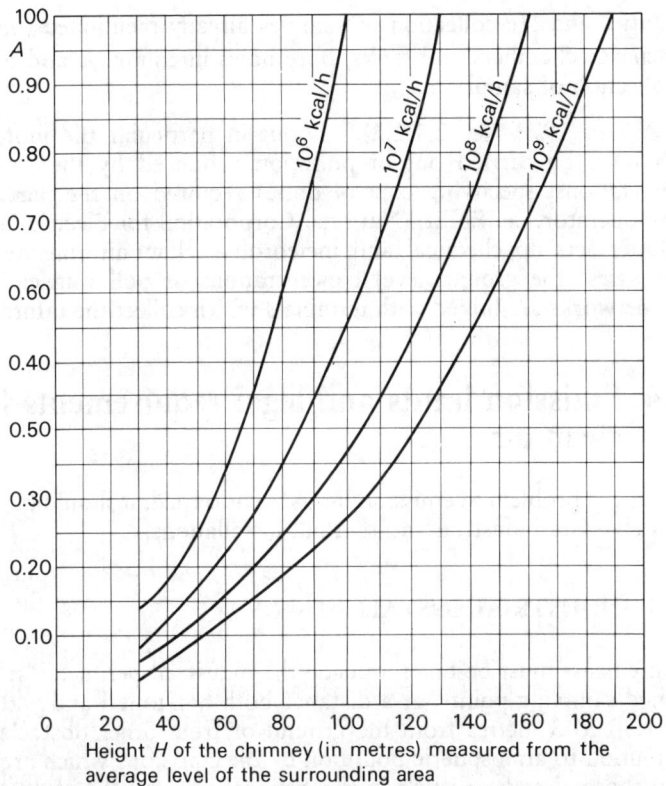

Permitted increases of solid particle content in smoke emitted by thermal installation in relation to height of outlet

2.2.3.3 SURVEILLANCE ACTIVITIES TO BE CARRIED OUT BY THE PARTY RESPONSIBLE FOR THE DISCHARGE

Many of the provisions mentioned with regard to the maintenance and correct operation of installations are combined with the controls which the party responsible for the discharge must carry out.

We should remember that the qualified operator of a thermal installation, in possession of a proficiency licence, can incur penal sanctions in the event of bad management (which involves considerable control duties), and that the operator of an industrial installation is obliged to report to the authorities considerable details of levels of emissions and periods and methods of operation of the anti-pollution equipment. We have also seen that DPR 322 of 1971 requires the use of reserve anti-pollution equipment in the event of a stoppage on the part of the principal equipment. The operator is responsible for checking the equipment (see 2.2.1) and in addition there are provisions relating to inspection duties and the collection of samples already mentioned (openings in discharge chambers and tanks, bore holes in chimneys and flues for the collection of samples etc.).

Law 880 of 1973 (see 2.1.2.2) sets out in particular the monitoring which must be carried out of pollution produced by thermoelectric power stations, specifying control duties required on the part of the station operator, i.e. ENEL (National Corporation for Electric Energy). The latter sets up chemical and meteorological monitoring networks which assess the ground level concentrations of polluting emissions. These networks are linked with terminals which collect the information.

2.2.4 Emission limits and legal requirements for clean air

As far as the problem of emission limits is concerned, it should be treated separately from industrial and thermal installations.

2.2.4.1 INDUSTRIAL INSTALLATIONS

Measurements must be taken outside the industrial perimeter at predetermined sampling points, at a distance both horizontal and vertical of from 1.50 to 3 metres from the ground or from other obstacles. The contribution to atmospheric pollution by the emissions which are identical to those already existing in the atmosphere, and therefore characterising the level of pollution already existing, can be measured either by simultaneous atmospheric readings, and readings taken from the source of emission itself, or by adding traces of substances to the emissions, or by using other suitable devices (article 9 of DPR 322 of 1971). See 2.2.6 for reference to the powers of the controlling authorities. These measures are aimed at ensuring that the limits set by DPR 322 article 8 as shown in the following table are not exceeded.

Law 880 of 1973 on thermoelectric power stations lowers the limits for SO_2 emissions (0.25 ppm for point concentrations and 0.10 ppm for daily average concentrations).

The values in the table can be modified by decree from the Ministry of Health in agreement with the Ministry of Industry, on advice from the Central Commission Against Atmospheric Pollution and the Superior Council of Health.

Some tolerance is allowed in the starting-up phase of installations which operate on a continuous cycle, but this is limited to 10% of the

STATIONARY SOURCES OF POLLUTION

Pollutant	Maximum point concentration 1013 millibar 25 °C ppm (mg/m³)	Duration of sample (minutes)	Frequency in 8 hours	Average concentration 1013 millibar 25 °C ppm (mg/m³)	Duration of sample (hours)
Sulphur oxides as SO₂	0.30 (0.79)	30	1	0.15 (0.39)	24
Chlorine (Cl₂)	0.20 (0.58)	30	1		
Chloric acid	0.20 (0.30)	30	1	0.03 (0.05)	24
Fluoride composites as fluoride	0.06	30	1	(0.02)	24
Hydrogen sulphur (hydrogen sulphide)	0.07 (0.10)	30	1	0.03 (0.04)	24
Total organic substances expressed as hexane Refinery derivatives	80.00	30	1	40.00	24
Nitrous oxide (NO₂)	0.30 (0.56)	30	1	0.10 (0.19)	24
Carbon oxide	50.00 (57.24)	30	1	20.00 (22.89)	8
Lead composite (Pb)	(0.05)	30	1	(0.01)	8
Inert suspended dust	(0.75)	120	1	(0.30)	24
Free crystalline silicate contained in dust as SiO₂	(0.10)	120	1	(0.02)	24

The pressure and temperature values refer to equivalent conditions among the concentrations expressed in volume amounts (ppm) and as weight per volume of air (mg/m³).

'Average concentrations' of pollutions are the amounts resulting from samples taken in a continuous manner and in constant amounts for the prescribed duration, or, in cases where the method of analysis will not allow lengthy sampling operations to be carried out, the average of the results of several shorter operations carried out successively over the prescribed period of time. 'Point concentrations' are the amounts resulting from samples taken continuously and in constant amounts for the prescribed duration. 'Frequency' is the number of times (established as in each 8-hour period) during which the emissions may reach the maximum amounts shown.

'Sulphur oxides' are the total amounts of SO₂ and SO₃ expressed as SO₂. 'Nitrous oxides' are the total amounts of NO and NO₂ expressed as NO₂. 'Inert suspended dust' includes all particles except those with a specific toxic action.

values in the table over a period of 6 months, with prior authorisation from the commune, on agreement with the Regional Committee. Similarly, the tolerance established by transitory orders of the regulation are in force for existing installations in communes which have just come within zone A or B: the limits in the table can be exceeded by a maximum of 50% in the first 2 years and 30% in the following 30 months (but these terms can be reduced by the communal authority). This provision applies from the date of the decree including the commune in zone A or B.

2.2.4.2 THERMAL INSTALLATIONS

Contrary to requirements for industrial installations, the prescribed limits for thermal installations refer to the emissions, i.e. to substances discharged in any way into the atmosphere (but not to immissions which can be recovered on the ground outside the industrial perimeter). The maximum limit for solid particles emitted is arrived at by the following calculation:

$q = 0.25(1+A)$

where q = the quantity of material emitted, in g/m^3, and A = percentage increase permitted for installations with an output of more than 1,000,000 kcal per hour by reason of capacity and height of outlet (see the table in 2.2.3.2);

i.e. the limit for installations with a capacity of less than 1,000,000 kcal/h is 0.25 g/m^3. The smoke index must not exceed no. 2 on the Ringelman Scale (but for installations using liquid fuel with chimneys over 50 m high it can reach no. 3 for not more than 5 minutes in each hour of operation; for installations using solid fuel, 10 minutes is permitted in similar circumstances).

The SO_2 content in the smoke should not exceed 2000 ppm (0.20% of the volume), this concentration to be checked at the base of the chimney.

The concentration of CO_2, measured at the base or outlet of the chimney, should be between 10% and 13% in volume for liquid fuel and over 10% for solid fuel to indicate good combustion.

The temperature of the smoke emitted should be not less than 90 °C at the point of entry into the atmosphere (it can also be measured at the base of the chimney and the heat loss during its passage to the outlet calculated).

These limits, according to article 13, clause 15 of DPR 1391 of 1970, are referred to in dry volume units at a temperature of 15 °C and a pressure of 760 mm of mercury.

2.2.4.3 LEGAL REQUIREMENTS FOR CLEAN AIR

The limits which we have already discussed do not in themselves constitute criteria for defining the quality standard for clean air from a legal viewpoint. A provision explicitly drawn up to this end does not exist; the only field in which the problem of environmental quality has emerged in the context of atmospheric contamination is perhaps that of the work environment which is a limited environment, the protection of which responds to the interest of a specific category of subjects.

However, Law 615 of 1966 states that its aim is to control emissions into the atmosphere of smoke, dust, gas and odours which alter the normal conditions of cleanliness of the air, and at the same time it also sets out parameters for assessing such alterations which can be either directly or indirectly harmful to the health of individuals and damaging to public and private rights.

This constitutes, by unanimous recognition, the beginning of direct legal protection of the conditions and quality of the air and it appears to some people, as already mentioned, that Law 615 already provides protection for the air as a distinct legal good.

2.2.5 Control of raw materials, fuels etc.

2.2.5.1 UNRESTRICTED FUELS

These are all the gaseous fuels (e.g. methane); petroleum distillates (kerosene, oil etc.) with a sulphur content of not more than 1.1%; gas and metallurgic coke containing not more than 2% volatile matter and not more than 1% sulphur; anthracite and anthracite products with not more than 13% volatile matter and not more than 2% sulphur; wood and wood charcoal (article 12 of Law 615 of 1966). It is considered that this list will be extended by a proposal from the Central Commission and a decree from the Ministry of Health, to cover products with similar properties resulting from technological advances.

2.2.5.2 FUELS WHICH CAN BE USED FREELY IN ZONE A BUT ARE RESTRICTED TO CERTAIN INSTALLATIONS IN ZONE B

These are combustible fluid oils with a viscosity up to 5 ° Engler and 50 °C and with a sulphur content of not more than 3%. In zone B these are permitted in industrial installations, and in thermal installations of more than 500,000 kcal/h.

AIR

2.2.5.3 FUELS PERMITTED IN BOTH ZONES ONLY BY COMMUNAL AUTHORISATION

These are:

combustible oils with a viscosity of more than 5 ° Engler and a sulphur content of not more than 4% (industrial installations and thermal installations of more than 1,000,000 kcal/h per thermal unit);

steam coal or charcoal containing volatile matter of not more than 23% and a sulphur content of not more than 1% (large, mechanically loaded boilers which use slow burning coal);

steam coal containing not more than 35% volatile matter and not more than 1% sulphur (same use as previous category but communal administration can, at its discretion, refuse to authorise the use of these fuels).

2.2.5.4 FUELS WHICH ARE PROHIBITED FOR CERTAIN USES

Agglomerates (e.g. briquettes, ovals) are in free use only for stoves for heating of premises. Lignites and peat are forbidden in zone B.

All fuels which do not conform to the specification laid down by the law for permitted use (limited to zones A and B to which the law is applicable) are forbidden. The implementing regulation contains tables specifying the characteristics of fuels in respect of their classification and distribution according to their permitted use. The tables are as follows.

Solid fuels

Fuel	Volatile matter (%)	Ash (%)	Sulphur (%)	Average size (mm)	Water content (%)
Metallurgical coke	2	8	1	over 40	8
		12		up to 40	12
Gas coke	2	8	1	over 40	10
		12		up to 40	14
Anthracite and anthracite products	13	10	2	all	5
Steam coal	23	12	1	all	6
Steam coal	35	12	1	all	6
Pitch lignite	40	20	10	over 40	5
				up to 40	10
Xyloid lignite	50	25	3	over 40	15
				up to 40	20
Peat lignite	40	30	2	—	25
Peat	40	30	2	—	35
Agglomerates	13	10	2	—	5

STATIONARY SOURCES OF POLLUTION

The data contained in the table are expressed in percentages by weight and represent the upper limits.
The average size, expressed in millimetres, indicates the average dimensions of single pieces of fuel contained in the deposits.
The percentages of volatile matter and ash refer to samples after complete drying.
The percentages of sulphur refer to fuel samples after drying to a constant weight, as specified in appendix 5 (of the law), and with the conventional 5% water content.
The water content percentages indicate the total content of water in the fuel samples taken.

Liquid fuels

				Fuel oil				
Characteristics	Limit	Unit of measurement	Gasoil	Very fluid	Fluid	Semi-fluid	Dense	Analysis
Opacity[a]	inf	mm	—	3	2	2	1	
Viscosity at 50 °C[b]	—	°E	—	less than 3	3–5	over 5–7	over 7	appendix 9
Water and sediment	sup	% in vol	0.05	0.5	1[c]	1[c]	2[d]	
Total sulphur	sup	% weight	1.10	2.5	3	4	4	appendix 8
Ash	sup	% weight	—	0.05	0.10	0.15	—	
Distillation at: 150 °C	sup	% in vol	2	—	—	—	—	
250 °C		% in vol	less than 65	less than 65	less than 65	less than 65		
350 °C		% in vol	85 or more	less than 85	less than 85	less than 85	less than 85	

To identify other petroleum distillate fuels not mentioned in the table, exempt from all use restrictions for the sole purpose of the prevention of atmospheric pollution, reference should be made to the classifications established in the provisions concerning the fiscal controls currently in force.

[a] The opacity should be total when observing the transparency of the fuel in a glass container of a thickness specified in the table, at a distance of 10 cm from an electric light with metal filaments and an output of 50 decimal candles.
[b] The correspondence between the viscosity shown in the table and cinematic viscosity is: 3 °E = 21.1 cSt; 5 °E = 37.4 cSt; 7 °E = 52.9 cSt.
[c] Measured in a centrifugal complex.
[d] Measured separately by extracting water (which must not exceed 1.5% in volume) and sediments (which must not be more than 0.5% in weight).

Directive EEC/716/75 concerning the harmonisation of the legislation of the member States relating to the sulphur content of certain liquid fuels is awaiting the promulgation of a law as delegated to the Government by Law 42 of 9 February 1982.

AIR

2.2.5.5 CONTROLS

Applications to the commune to authorise the use of fuels which are subject to the described restrictions must be suitably justified and documented. The mayor must obtain the opinion of the Provincial Fire Service and the Health Official. In the event of a negative decision, the applicant has the right of appeal to the Prefect within 30 days. The use of these fuels in a manner not conforming to the prescribed methods is punishable by a fine (which is doubled and falls on the supplier if the offence depends exclusively on the fuel and the responsibility of the supplier is proven).

Control of fuels is entrusted by Law 615 of 1966 (article 19) to the Provincial Fire Service which can request the collaboration of competent communal technical offices from the local authority.

The controlling instruments are contained in article 15, ff. of DPR 1391 of 1970. Article 15 describes the operation of extracting samples to be carried out by personnel from the Provincial Fire Service or communal technical offices, distinguishing between solid and liquid fuels, also with regard to the times and places for extracting the samples: solid fuel samples can be extracted from accumulations in coal bunkers or storage areas, during the unloading operation, or from the depots which serve the technical installation; liquid fuel samples can only be obtained from the special inlets in the service tanks of the installation. The results of these collection operations are reported verbally.

The fuel analysis is carried out by provincial health and preventive medicine laboratories or other laboratories authorised by the Ministry of Health to which the samples are sent. If the analysis reveals a violation of the regulations, the laboratory reports to the Provincial Medical Officer and communicates with the operator of the installation in question and with the Head of the Provincial Fire Service. The interested parties can ask for another analysis by a request to the Provincial Medical Officer. This further analysis is carried out by the Higher Institute of Health. If the outcome remains the same (or if the interested party has not requested a new analysis) the Provincial Medical Officer notifies the judiciary authority. Article 16 of the implementing regulation (DPR 1391 of 1970) describes the methods of analysis and in particular the analytical determinations to be carried out:

Solid fuels	Liquid fuels	
	Distillates	Fuel oil
Sulphur content	Sulphur content	Sulphur content
Moisture content		Viscosity
Volatile matter		

2.2.5.6 SPECIAL REGULATIONS

In consideration of the particular protection requirements of the Venetian lagoon, Law 171 of 1973 ('Law for Venice') makes the use of gaseous fuel compulsory for installations, even for those of less than 30,000 kcal/h and over 500,000 kcal/h.

Law 1083 of 6 December 1971 should also be mentioned since it contains safety provisions for the use of combustible gas and lays down suitable measures for avoiding dangerous situations arising from the malfunctioning of installations which run on such products.

2.2.6 Powers of the public authorities

In the course of this chapter we have already mentioned important powers, especially those concerned with authorisations required under anti-smog legislation and the control of fuels. We shall therefore not reconsider the requirements concerned with planning (see 2.1.2.1 and 2.1.2.2), with the classification of industries (2.2.1), with the granting of proficiency licences for the operation of thermal installations (2.2.2), with the authorisation for the operation of industrial installations equipped with anti-pollution equipment (2.2.1), or with the controls on fuels (2.2.5.6).

We shall, instead, discuss the powers of surveillance of the operation of installations and emission limits and their relative powers of sanction and constraint.

2.2.6.1 POWERS DERIVING FROM GENERAL PUBLIC HEALTH PROTECTION LEGISLATION (OPERATIONAL OVER THE WHOLE NATIONAL TERRITORY)

Under the provision of article 216 TU of the health law the mayor can prohibit the starting up of a factory included on the list of noxious industries (see 2.2.1) or he can make it subject to special precautions. He can also order, on the suggestion of the Health Official, the closure or relocation of a factory producing harmful emissions. The interested party can appeal to the Provincial Medical Officer within 30 days and this appeal results in a suspension of the mayoral provision.

According to article 217 TU the mayor must also prescribe the regulations to be applied to prevent damage or danger resulting from the emission of vapour, gas or other harmful matter arising from factories or processes and he can, in the event of non-compliance, officially intervene in the manner stated in the communal and provincial TU law. We should remember that article 153 of this TU empowers the mayor to issue contingency and emergency provisions in matters of public hygiene.

As regards the powers of the mayor in this matter, see: TAR Lombardia, 11 January 1978, no. 10, in *Trib. aum. reg.* 1978, I, 979; TAR Liguria, 4 May 1978, no. 207, *ivi* 2744; Cons. Stato., Section V, 14 April 1978, no. 451, in *Riv. amm.* 1978, 561; Cons. Stato., Section IV, 29 April 1977, no. 440, in *Giur. it.* 1978, III, 1, 214. It is often maintained in penal decisions that inactivity of the mayor in carrying out his surveillance and control duties in order to take suitable and effective measures fulfils the requirements of the punishable offence of failure to perform official duties (non-feasance) (article 328, Penal Code). Non-observance of the orders and official orders issued by the mayor under article 217 TU of the health laws constitutes a punishable offence under article 650 of the Penal Code (non-observance of orders from the authorities).

It should be remembered that the non-observance of obligations relating to the operation of a noxious industry exposes the offender to the penalty of a fine (article 216, final clause, TU health laws).

Directive EEC/312/77 concerning the biological monitoring of the population against the risk of lead poisoning is awaiting the promulgation of a law as delegated to the Government by Law 42 of 9 February 1982.

2.2.6.2 POWERS DERIVING FROM ANTI-SMOG LEGISLATION

The surveillance of thermal installations under article 19 of Law 615 of 1966 is the responsibility of the Provincial Fire Service by means of periodic checks or by notification from the health authorities or other controlling bodies specified by the law itself. We have already described the way in which these controls work and the relationship between the Fire Service, the Provincial Medical Officer and the other health authorities, when discussing the control of fuels (2.2.5.5). The last clause of article 19 states that

> the Commands of the Provincial Fire Service must notify the Communal Health Official, the Provincial Medical Officer and the Regional Committee, indicated under article 5, of all offences against the present law, the implementing regulation and local regulations encountered while carrying out the controls under the first clause of the present article, or with which they have, by whatever means, become acquainted.

The surveillance of industrial installations is the responsibility of the commune and the province, which can make use of information collected by the Regional Committee, possibly by means of on-the-spot investigations on the part of an appropriate Provincial Commission made up as follows: the Provincial Medical Officer, a representative of the commune, the Chief of the Provincial Fire Service, the Director of the Provincial Chemical Laboratory, an Inspector of Works, a representative of the Chamber of Commerce, a physico-chemical expert and an expert in industrial chemistry. If an installation does not conform to the required specifications, the commune can notify the interested party of his obligation to comply within a certain period, and in the event of non-compliance within this period, the offence is punishable by a fine. In addition, the Prefect can order the temporary closure of the installation. The competence of the Regional Committee, the mayor and the Prefect is replaced by those of the Central Commission and of the Ministry of Health in cases of pollution involving bordering communes from more than one region.

We have discussed penal sanction (fines) which can be applied under the anti-smog regulations, according to the offence committed (operation, construction, modification of installations without authorisation or testing, exceeding permitted emission limits etc.). The following list summarises the most important sanctions:

construction or alteration of a thermal installation without authorisation	fine of 100,000 to 1,000,000 lire
failure to notify authorities for the test	up to 50,000 lire
starting up a non-tested installation	up to 150,000 lire
use of prohibited fuels or improper use of permitted fuels	up to 300,000 lire
emission of smoke exceeding the permitted limits	up to 50,000 lire
non-compliance with requirements on the part of industrial companies or failure to respect the deadline fixed for the installations to conform	from 100,000 to 1,000,000 lire

These offences go together with those under article 650 of the Penal Code (non-observance of orders from the authorities). Law 171 of 1973 (Law for Venice) has increased the sanctions under Law 615 of 1966 tenfold.

See 2.1.1 on the relationship between the anti-smog regulation and punishable pollution offences. See 1.2.3.1 on the use of regulations under the Penal Code.

2.2.7 Rights of the individual

See 1.4 on rights to information, rights of appeal to judiciary or administrative authorities, rights to obtain the cessation of the harmful activity, and rights to compensation.

See 1.2.3.1 concerning the rights of individuals to make use of civil protection against immissions.

The anti-smog legislation also contains some methods of appeal to the administrative authorities on the part of individuals: any interested person can request the inclusion of a commune in zone A or B; private individuals, associations and public bodies can present observations and problems to the Central Commission Against Atmospheric Pollution and obtain an opinion; any interested person (not only the owner of a business) can contest the classification of industries into noxious or dangerous processes, and can appeal to the Provincial Medical Officer against the classification applied by the mayor if he believes it to be too comprehensive or alternatively if he considers it insufficient.

2.3 MOBILE SOURCES OF POLLUTION: AUTOMOBILES

Heading VI of Law 615 of 1966 is concerned with pollution from motor vehicles: the expression 'motor vehicles' includes motors designed as a means of transport and thus designated under the TU of the road traffic laws promulgated by DPR 393 of 1959 (i.e. mopeds, motor vehicles; trams and trolley buses however, since they are electrically propelled, do not come under Law 615; agricultural machines; trolleys and operating machines).

The provisions under heading VI apply to the whole of the national territory.

Law 615 of 1966 refers to the implementing regulation for the solution of many questions (particularly regarding emission limits). In addition, a regulation has been issued for diesel powered vehicles only, whereas Otto-cycle (or positive ignition) vehicles are covered under Law 437 of 3 June 1971, made following the need to adapt State legislation to EEC Directive 220/70.The provisions on diesel powered vehicles also apply to diesel powered trains belonging to the State railways (circular of 3 April 1973).

Subsequently EEC Directive 290 of 28 May 1974 intervened, allowing for the progressive technical updating of EEC Directive EEC/220/70 concerning the harmonisation of the legislation of the member States relating to the measures to be adopted against atmospheric pollution by gas produced by motor vehicles with positive ignition, following which the Ministerial Decree of 7 March 1975 provided for partial EEC type approval of motor vehicles with positive ignition with regard to emissions. Approval is granted by the relevant division of the Department of Civil Motorisation, following tests during which representatives from the Ministry of Health and the Superior Institute of Health can intervene. A further Directive concerning the same problem (EEC/102/77) has not yet been implemented.

2.3.1 Motor vehicles: controls for approval

Motor vehicles produced in Italy are subject to type approval by the Ministry of Transport; those produced abroad can request this type approval or alternatively they can undergo periodic checks to ensure that they conform with the requirements of the law; these checks are conducted by an engineer from the Department of Civil Motorisation.

The TU of the road traffic laws (DPR 393 of 1959) contains requirements for vehicle safety; Law 615 of 1966 adds requirements to limit the polluting capacity of the vehicle. There are two important principles: (a) vehicles must not produce polluting emanations exceeding the limits established by the implementing regulation; (b) the emanation of toxic or harmful products must be limited in cases of discharge (article 22 of Law 615 of 1966). The two principles go beyond the type approval controls, acting as directive criteria also for the maintenance and use of the vehicles. Nevertheless, they have an important application for type approval.

For motor vehicles with positive ignition, Law 437 of 1971 requires the vehicle to have at least four wheels, to be of an authorised weight, fully loaded, of not more than 400 kg, and to have a speed of at least 50 km/h; for type approval purposes, it must conform to the requirements laid down by the law. The necessary tests and analyses are aimed at establishing the levels of carbon oxide and of hydrocarbons emitted, in comparison with the prescribed limits, expressed in bulk for each test with respect to the reference weight of the vehicle (weight of the vehicle in working order, plus a fixed weight of 120 kg). Technical experts from the Ministry of Health and from the Higher Institute of Health participate in these tests conducted in accordance with procedures already established in the TU road traffic laws. Methods of collecting and analysing the gas can be authorised by decree from the Minister of Health with the agreement of the Ministers of Transport, Industry and Commerce and with the opinion of the Central Commission Against Atmospheric Pollution, as long as the results are of the same value.

As regards diesel powered motor vehicles, Law 615 of 1966 contains a control on the opacity of exhaust emitted, establishing this as the parameter with which levels of toxic or harmful emanations are compared. The implementing regulation (DPR 323 of 1971) states that the opacity standards of the fumes must not exceed 45% for urban buses and 50% for all other vehicles in order to obtain type approval for road use. This opacity is measured by ascertaining the percentage of the flow of white light absorbed by a column of exhaust 40 cm wide (value 0 = complete transparency). Article 3 of DPR 323 describes in detail the measuring operations, also specifying the environmental conditions (not more than 800 m above sea level). The measuring apparatus (opaci-meters) must be of a type approved by the Ministry of Transport and must also be able to carry out a continuous measurement with evidence of transitory phenomena.

Under penal and civil law the manufacturer is responsible for ensuring that the vehicle produced conforms with the prototype which has received ministerial approval, formal declaration being given for each finished vehicle produced.

There are no regulations in existence making motorcycles or motor vehicles subject to type approval, nor are there controls relating to their polluting potential.

Article 22 of Law 615 of 1966 enables the Minister of Health to issue ministerial decrees in agreement with the Ministers of the Interior, Transport, Industry and Labour, and having heard from the Superior Health Commission, making the use of efficient applicances to reduce the toxicity of exhaust fumes obligatory for diesel powered vehicles and Otto-cycle vehicles.

Directive EEC/306/72 relating to the measure to be adopted against pollution produced by diesel engines intended for the propulsion of vehicles was implemented in Italy by ministerial decree on 5 August 1974. However Directive EEC/537/77 on the same subject is still awaiting implementation.

2.3.2 Motor vehicles: controls over maintenance and use

Article 55 of the TU of 1959 relating to road traffic provides for regular or special inspection visits to check the state of maintenance of vehicles.

The Ministry of Transport organises these regular checks either by ministerial decree at least every 5 years (general or partial inspection of all motor vehicles and side cars for private use) or by subjecting the vehicles to annual tests (compulsory for all motor vehicles not referred to above).

Article 24 of Law 615 of 1966 states that during the course of these inspections it should also be established that the motor vehicles do not produce polluting emissions.

Special inspections can be ordered at any time for single vehicles by the Inspectorate of Civil Motorisation when there is reason to believe that a vehicle does not conform to the requirements of the law.

In this respect, Law 615 of 1966 has established that the driver of a diesel powered vehicle which emits fumes with an opacity exceeding the limit set by the regulation, besides being liable to a fine (see 2.3.4), is obliged to eliminate the cause of such immissions, and that compliance with this obligation must be verified by submitting the vehicle to inspection by an Inspectorate (Management Office) of the Civil Motorisation Department, or by a detached office. Likewise, when there is reason to believe that the immissions from a motor vehicle do not comply with the requirements of the law, the vehicle must be submitted to an inspection similar to that required under clause 3 of article 55 TU on road traffic (special inspection).

The opacity values permitted by the regulation for emissions from diesel powered engines already in circulation are higher than those permitted by type approval, i.e. 65% for urban buses and 70% for other vehicles.

For Otto-cycle vehicles the limits under Law 437 of 1971 are only valid by type approval and are not applicable to vehicles already used on the road.

There are no specific requirements relating to the protection of the atmosphere from pollution as far as the use of motor cars is concerned. Nevertheless, it is fairly common for the communal authorities to close important areas of the city to road traffic (historic city centres etc.). The application of such provisions, principally due to considerations of an urban planning or sociological nature, is also being extended to aspects of the restriction of harmful emissions: for example, experimental projects involving the closure of city zones to road traffic to avoid the corrosive effect of exhaust fumes on monuments or buildings of historic or artistic merit.

2.3.3 Fuel

The anti-smog legislation does not contain requirements concerning fuels which may be used. Nevertheless, the use of fuels by motor vehicles is controlled by tables approved by the Ministry of Transport on the basis of the regulation promulgated by DPR 420 of 30 June 1959.

2.3.4 Powers of the public authorities

See 2.3.1 for powers of type approval, and 2.3.2 for inspection.

It has been observed that the general requirements under article 22 of Law 615 of 1966 are not provided with sanctions in the event of violation. Only the driving of a diesel powered vehicle emitting fumes exceeding the permitted opacity limits is punishable with a fine of 5000 to 20,000 lire. This is a very light punishment, which can in practice be as little as 7000 lire. In the diesel powered motor section the requirements of the law and the regulations are often ineffective as a result of the extreme complexity of the verifications during type approval and the shortage of personnel and the scarcity of control structure.

The situation relating to motor vehicles with positive ignition is even worse. As we have already seen, vehicles in this category are not subject to emission limits: Law 436 of 1971 states limits only for type approval purposes, while the most dangerous emissions almost always occur after type approval; even if emission limits did exist, exceeding them would not be a punishable offence.

In order to avoid these problems, it has been proposed that article 674 of the Penal Code (see 2.1.1 and 1.2.3.1) should be applied. Under this article anyone who causes offensive, dirty or troublesome emissions of gas, vapour or fumes, *in cases not permitted by law*, is liable to up to 1 month in prison or a fine of up to 80,000 lire.

In preparing penal sanctions, this provision could integrate the principles of article 22 of Law 615 of 1966 by considering emissions of toxic or harmful products on the part of motor vehicles to be 'cases not permitted by law'.

Finally, article 23 of Law 615 of 1966 makes the verification of such violations the responsibility of civil servants, officials and police officers under article TU of the road traffic laws (Highway Police, Carabinieri, Customs and Excise Officers, Municipal Police, etc.).

2.3.5 Rights of the individual

There are no specific regulations: this area is covered by the more general provisions described in 2.2.7 and 1.4.

2.4 SHIPS AND AIRCRAFT

2.4.1 Ships

There are only individual provisions relating to ships. Law 171 of 1973 (Law for Venice) has extended the application of the laws relating to atmospheric pollution by motor vehicles to barges on the Venetian lagoon with effect from 8 May 1975; Law 59 Prov. Trento of 1973 has prohibited barges with diesel powered and Otto-cycle engines to be used on State-owned lakes in the province with a total surface area of less than 1 km^2; in the other lakes of the province only motors with an engine capacity of more than 4 HP are permitted.

2.4.2 Aircraft

In spite of the belief that Law 615 of 1966 also extends to aircraft, it must be considered that this source of pollution is not covered by the law as it is not explicitly included in the statute.

The work of the Italian Aeronautical Register (RAI) is significant in this context since it has important attributions in the planning and construction stages of aero-engines (granting of type approval and airworthiness certificates, approval of foreign type approvals etc.). The RAI technical provisions are of particular relevance, which contain the technical requirements covering aero-engines (these do not include requirements to restrict the polluting potential of engines). The RAI also authorises possible modifications and promotes controlling inspections to check the condition of aircraft in use.

But the surveillance and overhaul of aero-engines to restrict toxic and harmful emissions is almost exclusively the responsibility of the airline companies themselves.

The provisions which can be applied in relation to pollution by aero-engines are article 674 of the Penal Code (see 2.1.1 and 1.2.3.1) and article 819 of the Navigational Code, which prohibits the jettisoning of objects or matter from aircraft without the specific authorisation of the Ministry of Transport. The Navigational Code makes it compulsory for the manager of aircraft to be insured against injury to third parties on the ground: any third party sustaining injury can apply directly to the insurer for compensation.

2.5 REGIONAL LAWS AND OTHER SPECIAL PROVISIONS

As regards regions with special statutes, there is extensive legislation in Trentino-Alto Adige: Law 59 Prov. Trento of 1973 has already been mentioned (2.4.1); there is also a provincial law (No. 24 of 30 August 1968) which institutes credit facilities to favour equipment to eliminate harmful fumes in industries.

There are also two important laws in Bolzano province which should be discussed (No. 12 of 4 June 1973, and No. 46 of 13 September 1973), the second of which completes and modifies the first. As a result of this provincial legislation, exceptions are in fact made to the requirement of the TU of the health laws and Law 615 of 1966 in the territory of the province. The laws in question apply to both industrial and civil thermal installations with a capacity of more than 30,000 kcal/h. The anti-pollution equipment must be approved at the time building permission is granted and such equipment can only be permitted to start functioning if it has been tested. Requirements like those under Law 615 of 1966 are also in force for motor vehicles. The surveillance and control authorities are the mayor, the Provincial Office, the Provincial Assessor of Health, and the President of the Provincial Council (who can order the closure of installations which do not comply with the requirements). The law also contains provisions for the prevention of air pollution in buildings and places of work.

The regions with ordinary statutes have to date legislated almost exclusively to provide for the implementation of measures relating to the constitution and functions of the Regional Committees Against Atmospheric Pollution (notably Lombardy since 1972 and the other regions between 1974 and 1975).

SPECIAL PROVISIONS

The remaining legislation frequently corresponds to particular contingencies and emergency situations (see the laws of the Lombardy region on the occasion of the ecological disaster at Seveso following an unexpected breakdown at the ICMESA plant).

As far as special State laws relating to the subject in question are concerned, the Law for Venice of 1973 has already been mentioned but Law 584 of 11 November 1975 should also be included. It prohibits smoking in specific public places (cinemas, theatres, schools etc.) and on public transport. Offenders are fined under administrative proceedings.

3
Inland Waters and Discharges into Public Sewers

The legislation on water protection from pollution has culminated in Law 319 of 1976 (known as 'Merli Law') which made provision for a series of instruments for State and Regional intervention (delegating many powers to the latter). This law has proved largely ineffective, mainly due to lack of finance as a result of the failure of the self-financing mechanisms envisaged by the law itself.

For this reason a new law was made necessary: Law 650 of 24 December 1979 (known as 'Merli bis').

We are, therefore, in the middle of a transitional phase limited, however, to the legislation since, on a practical level, the provisions contained in Law 319 of 1976 have had very little effect. In fact, the need for further legislative intervention was forseen from the moment that Law 319 of 1976 was issued, since the obstacles and resistance which the law encountered led to a restriction of the contents resulting in partial and substantially superficial provisions rather than the comprehensive redefinition of the water provisions which had been the original intention behind the law. In fact, a preliminary draft of the law produced by a Government commission was decisively rejected as a result of opposition from the Northern Regions which found support inside Parliament. The Lombardy region, in particular, which during the course of these events had planned and issued its own Regional Law 48 of 19 August 1974 which was proving very detailed in its control of water pollution, was strongly opposed to any solution which did not enhance the role of local autonomy, not only for pollution protection purposes, but also in the reorganisation of the water regulations (water management, a completely reorganised administration for water with criteria of autonomy and decentralisation, the attribution to local bodies of active administrative functions and of a planned organisation of intervention). But the Northern Regions, the front runners in the environmental sector, did not succeed in obtaining such a vast reform, primarily because they did not have the support of the Southern Regions, whose internal administrative organisation and different environmental situation led them to prefer a more centralised solution. In order to avoid a complete standstill, the scope of the law was limited to the problem of polluting discharges; the question of planning and administrative reorganisation was deferred. In technical content the law substantially conforms to the draft suggested by the regions at the end of their heated discussion and it was very similar to Law 48 Reg. Lomb. of 1974 on which it was modelled. It did not, however, represent a great step forward for regions such as Lombardy and Piedmont which had already made their own provisions independently, but for a large majority of the other regions it was the source of a great number of new attributions.

Subsequent regulatory intervention has made some of the provisions of Law 319 of 1976 ineffective (e.g. Law 690 of 1976 which reduced the areas of effectiveness of the provisions on manufacturing installations), or has extended the period for compliance with provisions under Law 319 (e.g. Law 690 of 1976 concerning the purification of discharges and the payment of sums provided for under article 18 of Law 319 of 1976 by manufacturing installations as compensation for damage caused by the discharge; the presentation of applications for authorisation for the discharge, etc.).

The new Law 650 of 24 December 1979 has not replaced Law 319 of 1976 but has corrected it and integrated the points in which it was inadequate. In particular, the financing mechanisms for intervention provided for by the law have been revised. Nevertheless, the 'Merli bis' law introduces important innovations even with regard to essential aspects of the control of discharges, by altering the distribution of function and powers of the various administrations involved (with a marked increase in the sphere of regional authority at the expense of the province and of the peripheral administration of the State). It also amends the system of limits and requirements with which discharges must comply.

3.1 CONTROL BY LAND-USE PLANNING

Law 319 of 1976 does not directly formulate a plan for the control or treatment of water but provides the State, the regions and other local bodies with various attributions constituting a combination of planning instruments and activities.

Under article 2 of Law 319 of 1976 the State is responsible for drawing up the general plan for water treatment on the basis of the regional plans. It is also responsible for controlling the compatibility of these regional plans through permanent inter-regional conferences promoted by the Minister for Public Works with regard to river basins involving several regions.

The regional plans for water treatment are the responsibility of the individual regions and, according to article 8 of Law 319, had to be sent to the Committee of Ministers (a body set up by the same law which, since Law 650 of 1979, is known as the Interministerial Committee), within 3 years of the entry into force of the law and had to be carried out within 10 years of that date.

Today, under Law 650 of 1979, the regions must prepare a preliminary programme for water treatment (concerning the fundamental objectives and priorities of work) and send it to the Interministerial Committee before 31 March 1980.

The law defines the contents of the regional plans—reorganisation of the peripheral technical and administrative structure for the following services: aqueducts, sewerage, purification; programming of related public works; definition of criteria and deadline for intervention.

The principal defect of this planning system is the lack of integration of the regional water treatment plan and urban planning (regional coordination of the territorial plan, see 1.2.2.2). In short, the regional territorial plan performs a purely urban planning role despite the efforts of the regions to extend it to include environmental problems, and Law 319 has not filled this gap between urban planning matters of regional responsibility and environmental matters. It should have been possible to unite the regional treatment plan and the territorial plan in one comprehensive planning process; failure to do so has resulted in a regional treatment plan which is not comprehensive with regard to its functioning mechanisms (there is no provision for a legal or administrative instrument to organise the plan, nor are there any procedural provisions concerning State and regional intervention).

There are, however, other regional planning possibilities under Law 319, such as the regional provision integrating and implementing the general criteria for the correct use of water (article 4 of Law 319 of 1976) which can act as a planning instrument for various problems (disposal of sludge and effluent, application of technical requirements and general planning criteria for water use etc.).

Amongst its varied planning duties, the region must also coordinate the programmes prepared by the various local bodies. The nature and extent of these programmes is not, however, defined. Article 8 of Law 319 of 1976 states that the entire regional treatment plan must be proposed after hearing from the communes involved, thus creating a framework for intervention by lesser local bodies in the detail of the plan. These local bodies must have responsibility for matters listed under article 5 of Law 319 (in particular, control of the application of general criteria for the correct and rational use of water) and article 6 (public services for aqueducts, sewerage, purification, disposal of sludge etc., control of discharges for limits of acceptability, installation and maintenance of the network of appliances for quality control of the waters).

The census of water sources required under Law 319 of 1976 is one of the basic instruments for the purpose of planning (and not only for that purpose); the regions are responsible for carrying out this census and for sending the resulting information to the Committee of Ministers for the compilation of the national treatment plan. The information to be collected concerns: (a) the hydrological, physical, chemical and biological characteristics of the water sources and their flow; (b) all the direct and indirect uses: utilisation, derivations, discharges. The law allowed a period of 2 years in which to carry out the census and a similar period for subsequent updating, but this has been extended to 31 March 1981 (Law 650 of 1979). A further extension of one year has subsequently been granted by DL 801 of 30 December 1981. There are therefore two distinct requirements: to obtain precise information on existing water sources through the collection and computation of data, and to institute a proper register not only of discharges but of all water uses, in accordance with the requirements of the law (a comprehensive control of the water cycle). Article 7 of Law 319 completes the latter requirements by sanctioning the obligation on the part of private individuals who have their own water supply (with the use of suitable instruments which they themselves must procure) to measure the amount of water extracted and to declare this information to the relevant authorities (Provincial, Communal or Consortia) at most once a year. Automatic measuring instruments can also be imposed by the authorities.

The fact already underlined, that the law is not only concerned with discharges but also with the entire cycle of water utilisation, merits examination in the light of what has been decided for the correct and rational use of water resources.

3.2 CONTROLS OVER THE USE OF WATER

Article 2 (*d*) of Law 319 of 1976 leaves to the State

> the indication of general criteria for a correct and rational use of water for manufacturing purposes and industrial and civil uses also

by means of the identification of standards for consumption to favour maximum savings in the use of water....

This State function is carried out by the Interministerial Committee created by article 3 of Law 319.

On the same subject the regions have legislative competence for the enactment of laws (article 4) and the provinces the power of ensuring that such legislation is observed.

The Interministerial Committee has laid down the general criteria for the correct and rational use of water in its resolution of 4 February 1977, annex 2. These criteria refer to nearly all water uses and act as planning guidelines for the use of water resources.

In this sense, the definition of the general criteria under discussion has a significance which goes beyond the problem of water pollution and covers the question of the most convenient distribution of utilisable resources. In this way the activity of the State, and regional provisions for coordinating the way in which this activity is carried out, can be seen as a planning instrument at regional level and, as underlined above, this regulatory power which, in this sector, is delegated to the regions indicates a subtle extension of the scope of regional planning activities which is otherwise confined within the limits of the regional treatment plans. In this respect it has been noted that

> if alterations to the use of water can have an influence over discharges, the most intelligent policy does not consist of purifying the discharges ... but in modifying them in order to improve their characteristics, with intervention on the causes [DI FIDIO]

thus aiming at achieving the best possible ecological conditions through planning of water resource use rather than trusting to absolute conformity with the limits contained in the table of permitted discharge limits.

EEC Directives EEC/160/76 concerning the quality of bathing waters, EEC/440/75 concerning the quality of surface waters intended for drinking water use in Member States and EEC/659/78 on the quality of fresh water requiring protection or improvement to be suitable for fish life, are awaiting the issuing of laws due as a result of the powers delegated to the Government by Law 42 of 9 February 1982.

3.3 CONTROL OF DISCHARGES INTO WATERCOURSES

Law 319 of 1976 controls discharges arising from public sewers and municipal and manufacturing installations throughout the whole national territory. The problem of discriminating between civil and manufacturing installations for the purposes of applying different regulations to each category has found a less than effective solution in Law 690 of 1976.

This law has introduced a very restrictive definition of manufacturing establishments, thereby extending the effective range of control over civil plants: the term 'manufacturing establishment' means one or more buildings or installations linked together in a determinate area from which one or more terminal discharges originate and in which the prevailing activity is the production of goods in a permanent and stable manner.

Under the new Law 650 of 1979 the Interministerial Committee, including the Minister for Agriculture and in consultation with the regions and professional agricultural organisations of national importance, is responsible for the definition of agricultural enterprises which should be considered as civil installations within the meaning of Law 690 of 1976 (this definition was to be carried out within 60 days of the promulgation of the law).

The main consequence of the classification of a discharge as arising from a civil or manufacturing installation concerns the permitted limits (see 3.5); to this end, measurements are taken immediately above the point of discharge into the receiving watercourse.

EEC Directive EEC/439/75 regarding the disposal of used oils is awaiting the issuing of a law due as a result of the powers delegated to the Government by Law 42 of 9 February 1982.

3.3.1 Authorisation for the discharge

Article 9, final clause, of Law 319 of 1976 requires that 'all discharges must be authorised' and adds that the authorisation must be granted by the relevant controlling authority. Subsequent to Law 650 of 1979 and with reference to discharges into surface watercourses or public sewers only, this authority is bestowed on the individual or associated communes and mountain communities.

The enlargement, restructuring or change of use of civil or manufacturing installations producing discharges also requires authorisation (article 10 of Law 319).

The authorisation must conform with the permitted limits provided for by the law (see 3.5.1). For existing discharges a distinction must be made between civil installations which do not discharge into public sewers, and manufacturing installations. The person responsible for discharges arising from the former is obliged to declare his position to the communal authority; the person responsible for discharges from the latter must either apply for authorisation for the discharge or apply for renewal of the authorisation already obtained. Article 15 of Law 319 laid down various time limits for the carrying out of these requirements: 2 months from the entry into force of the law for the authorisation application and 6 months for renewal of the application and for the declaration to the communal authorities regarding discharges from civil installations which do not enter public sewers, the deadlines to be decided by the communal authority. The authorisation for the discharge from civil installations into public sewers is not bound to comply with the permitted limits, but with the regulations issued by the local authorities (communes, intercommunal consortia).

The law refers to the regional treatment plans for the control of discharges from civil installations which do not go into public sewers (and this is one case in which the person responsible for the discharge is obliged to declare it if it was already in existence when the law was issued, particularly since, while waiting for the regional plans to be prepared, the communes and communal consortia have to develop plans for construction of the drainage network).

Law 650 of 1979 states that the region must take into account the directives fixed by the Interministerial Committee and promulgated on 30 December 1980 (G.U. 10 January 1981) in issuing regulations for these discharges. But even before the plan was promulgated the regions provided the necessary measures for the protection of public health through their health organisations (article 17 of Law 650).

Article 15 of Law 319 establishes the data and information which must be supplied in order to obtain the authorisation or renewal: qualitative and quantitative characteristics of the discharge, quantity of water extracted in a calendar year, possible alternative destination for the discharge, supply source. Two forms of authorisation are provided for: provisional and definitive. The latter is subject to compliance with the permitted limits fixed by the law.

On the other hand, provisional authorisation provides for a progressive alignment with the limits established in the various tables attached to the law according to the water source which receives the discharge (discharges into public sewers must comply with the permitted limits and technical requirements laid down by the commune or consortium managing the water service, whichever is appointed under the regional treatment plan).

Authorisation is understood to be granted if there is no response from the administration concerned for a period of 6 months after the application but it can be subsequently revoked or be subject to compliance with certain requirements. The power to revoke the authorisation also exists in cases of non-compliance with the permitted limits.

Law 650 of 1979 has completed article 15 of Law 319 of 1976, with the requirement that costs incurred for verifications and inspections required for the preliminaries relating to the authorisation application should be charged to the applicant (a sum has to be paid for this purpose before the application can be processed).

Apart from the general requirements for the authorisation, the matter appears to be controlled in a very confused and incomplete manner: Is there any obligation to declare discharges from manufacturing plants which do not go into public sewers? Can the people responsible for effluent from existing manufacturing installations discharging into the ground or subsoil present an authorisation application to the commune as long as no intercommunal consortia has been constituted? There are many questions of this kind created as a result of the authorising regime provided for in the law.

Article 2 of Law 650 of 1979 has introduced a further authorisation procedure (not for discharges but for purification plants installed to ensure conformity with the permitted limits) for manufacturing installations already in existence when Law 319 of 1976 was promulgated and which have not yet complied with the limits set down by the law with regard to discharges. Within 2 months from the entry into force of Law 650 of 1979 those responsible for these discharges must present a detailed programme showing the date of commencement of the work, the amount of time it will take and the cost. Within 3 months the regions must authorise the commencement of the programme with the inclusion of further requirements if necessary. The deadline for the completion of this work is 1 September 1981, extendable by one year, as established by Order in Council (Decreto Legge-DL) 801 of 30 December 1981. The authorisation can be revoked if the programme is not carried out.

3.3.2 Powers of inspection, monitoring and control

We have seen that the law determines the measurement criteria for permitted limits for discharges, establishing that measurement must be taken immediately above the point of discharge. For this reason the discharge must be accessible for samples to be taken at 'the point chosen for measurement' (article 9 of Law 319 of 1976). This applies to all discharges. For those arising from manufacturing installations it is also stated that the competent 'controlling authority is authorised to carry out all necessary inspections inside the manufacturing installation to check the conditions which give rise to the formation of the discharges'.

It has already been pointed out that the control of discharges into surface watercourses and discharges from manufacturing installations into public sewers is the responsibility of the communes and communal consortia (as is also the control of discharges into the ground or subsoil).

According to article 15, clause 6 of Law 319 of 1976 (as amended by Law 650 of 1976) the technical monitoring and control over all discharges is carried out by the special multizonal surveillance centres and multizonal services for the control and protection of environmental hygiene (Law 833 of 23 December 1978) (see 1.3.1). The general technical provisions for the regulation of purification plants also require the plant manager to make periodic checks on the hydraulic, chemical, physical and biological characteristics of the effluent both before and after treatment; these data must be entered, at the request of the controlling authority, into suitable notebooks which must be kept available at the plant for inspection.

3.3.3 Duties and obligations on the part of the discharger

In addition to the obligations of declaration, application for authorisation or renewal etc. already mentioned, there are other requirements with which the discharger must comply. Article 9 of Law 319 of 1976 states that the permitted limits for the discharge cannot be achieved by the dilution of the discharge with water extracted exclusively for that purpose. It is thus implicitly admitted that there is a possibility of conveying, within the same discharge, waters arising from different uses within the same plant (different processes, sanitation etc.). The prohibition on dilution of discharges in order to reduce the concentration of pollutants could also be evaded by the mixing of dangerous discharges with other water containing low concentrations of pollution, perhaps by extracting a greater quantity of water than that actually required. However, according to article 9, clause 5 of Law 319 of 1976, following an inspection inside the installation (3.3.2), the authority can request that partial discharges (before they are combined with the general discharge) containing particularly dangerous substances (e.g. toxic metals which have a tendency to accumulate inside living organisms) are subjected to special treatment before they are combined with the general discharge.

In such cases the objective is not only to avoid high concentrations of these substances but also to reduce, or reduce to zero, the quantity of toxic substances discharged into watercourses during determined periods of time by containing them instead at the bottom of the partial discharge which contains them. Perplexity arises concerning the nature of this power of the controlling authorities which is exercised only optionally (the regional provision which integrates and implements the general criteria for the correct and rational use of water—article 4 (c) of Law 319 of 1976—could propose a remedy for this inadequacy).

The prohibition on dilution of discharges (even partial discharges and even with cooling or rinsing waters which were not extracted solely for dilution purposes) has been extended under Law 650 of 1979 to cover partial discharges which must undergo treatment to bring them within the limits of the law.

Article 25 of Law 319 has also established a transitional regulation on the subject (valid until the permitted limits as laid down in the law have been observed), obliging the discharger to avoid even temporary increases in pollution (not an increase in the pollution concentration in the receiving water source but an increase in the polluting content of the discharge itself).

The new text of article 9 also requires the controlling authority to set discharge regulations (on the basis of the nature of the deterioration and on the objectives for protecting the water sources set by the regions) when the water extracted from a surface water source presents characteristics exceeding the tabulated limits. In any case, water must be replaced in the same water source in the same qualitative and quantitative conditions as it was extracted.

The discharger is also subject to financial obligations. Article 16 of Law 319 of 1976 requires the payment of a water rate to the communes and intercommunal consortia as a contribution towards the water services of abstraction, transport, purification and discharge. The compilation of lists of names for the collection of the water rates is the responsibility of the regions (article 17 of Law 319). There is, however, no similar provision for regional responsibility under article 18 regarding the payment of a sum in partial compensation for damage caused by discharges arising from manufacturing installations; this sum is calculated in direct proportion to the quantity and quality of water replaced. The sum in question is collected by the communes and the consortia which will be carrying out the anti-pollution works with it and it is payable from the date of the entry into force of the law, up to the date when the public and private devices provided for in the law, for the purposes of achieving the final water treatment objectives, are put into operation. This self-financing mechanism for anti-pollution initiatives has been hindered by the damaging amendments made to Law 319 of 1976 by Law 690 of 1976, restricting the application of the provisions in question both by narrowing the definition of manufacturing installations, and by introducing an exception to article 18 of Law 319 in altering the starting date for the payment of the compensation tax (which is no longer the date of the entry into force of Law 319, but the date of the resolution by the Committee of Ministers concerning the definition of the ground rent: 13 May 1977). In addition, while determining these criteria, the Committee of Ministers has established that the sums to be paid should be reduced by one-tenth per year for reasons of economic crisis. This operational failure on the part of the self-financing mechanism has resulted in the need for a basic reform of the control of discharges which has been largely carried out by the new law (Law 650 of 1979). This law has tripled the tax under article 18, payable by all dischargers who do not comply with the limits laid down therein by 13 June 1979; this deadline was extended to 1 September 1981 by Order in Council (D.L.) 801 of 30 December 1981. The sums collected in this way, together with other payments deriving from article 18 which are accounted for separately, are used to carry out the services under Law 319 of 1976. The compilation of the lists of names for the collection of this tax must be completed by 1 November 1980.

3.4 CONTROL OVER DISCHARGES INTO PUBLIC SEWERS

As already mentioned, Law 319 of 1976 covers not only discharges into watercourses, land and subsoil but also effluent discharged into public sewers, which should be discussed separately. The public sewerage and water purification services are managed by communes and intercommunal consortia, by mountain communities or consortia set up by the regions with a special statute, or by consortia for the areas or industrial development centres (nuclei) which are also responsible for the control of discharges into the public sewerage system from manufacturing complexes (acceptability of discharges, pretreatment plants, conformity with general criteria for the correct and rational use of water). Furthermore, the regional treatment plan should provide for the reorganisation of the peripheral technical and administrative structures responsible for aqueducts, sewerage and purification services as well as for the programming of relevant public works. To this end the law requires communes and consortia to provide the regions with programmes for setting up the sewerage networks they have planned. It is the responsibility of the communes or consortia to issue regulations concerning discharges into public sewers arising from civil installations (these are only obliged to comply with the regulations and not the permitted limits laid down by the law: see 3.5.2). See 3.5.2 regarding the permitted limits for discharges into public sewers. The law, however, only establishes limits for the period before the activation of the centralised purification plant which constitutes the main object of the programmes of the communes and consortia coordinated with the regional treatment plan. Once the objectives of the regional plan have been achieved, the acceptability of the discharges will be measured against the provisions and requirements issued by the communes and consortia managing the public sewerage service. Besides, in amending articles 12 and 13 of Law 319 of 1976, Law 650 of 1979 establishes that even before the central purification plants begin functioning, the communes and consortia can lay down more restrictive limits and requirements than those under the law, approved by the region, taking into account the local conditions concerned. Conformity with these requirements and limits must be achieved within 90 days from regional approval. If the purification plant does not begin operating by 31 December 1981, the legal limits come back into force.

We have already said (3.3.3) that communes and consortia receive a water rate from the discharger. This water rate consists of two sums, one for the sewerage service and one for the purification service.

The two laws mentioned here largely implement the EEC directive concerning pollution produced by certain dangerous substances discharged into the waters of the Community (EEC/464/76).

3.5 EMISSION AND CONCENTRATION LIMITS

The law talks of permitted limits for discharges, determined by requirements concerning the quality of waste waters or of the waters in the receiving water source, or alternatively by concentration limits for certain substances. The limits vary according to the destination of the discharges and, in the event of discharges into the same receiving water sources, according to the origin of the discharge (civil or manufacturing installation).

3.5.1 Discharges into surface watercourses

Discharges from new manufacturing installations must conform with the permitted limits in Table A in the law from the moment these installations begin operating; the discharge from existing manufacturing installations must conform to the same limits within 9 years of the entry into force of Law 319 of 1976, according to the methods and timetable established in the regional treatment plans.

Table A

No.	Parameter	Concentration	Notes
1	pH	5.5–9.5	The pH value of the receiving watercourse must be between 6.5 and 8.5 in a 50 m range from the discharge.
2	Temperature °C	—	For watercourses the maximum variation between the average temperature of any section of the watercourse above and below the discharge point must not exceed 3 °C. On at least half of any section below the discharge point the variation must not exceed 1 °C. For lakes the temperature of the discharge must not exceed 30 °C and the increase in temperature of the receiving water source must in no case exceed 3 °C at more than 50 m from the discharge point. For canals the maximum average temperature of the canal water at any point below the point of discharge must not exceed 35 °C. This is subject to the approval of the authorities in charge of managing the canal. For the sea the temperature of the discharge must not exceed 35 °C and the increase in temperature of the receiving source must in no case exceed 3 °C at a distance of more than 1000 m from the point of discharge. In addition, the formation of thermal barriers at river mouths must be avoided.
3	Colour		Not perceptible after dilution 1 : 20 over a width of 10 cm.
4	Odour		Should not cause inconvenience or nuisance of any kind.
5	Coarse materials	Absent	'Coarse materials' refers to objects of any nature with a linear dimension greater than 1 cm.
6	Sedimentary materials (ml/l)	0.5	Sedimentary materials are measured in an Imhoff cone after 2 hours.
7	Total materials in suspension (mg/1)	80	'Total materials in suspension', regardless of their nature, refers to materials of a size which does not allow them to pass through a filter membrane of 0.45 μm porosity.
8	BOD_5 (mg/l)	40	For industrial discharges whose oxidising characteristics are different from domestic effluent, the concentration limit should be referred to at least 70% of the total BOD.

Table A continued

No.	Parameter	Concentration	Notes
9	COD (mg/l)	160	The COD should be determined with boiling bichromate of potassium after 2 hours.
10	Total toxic metals and non-metals (As-Cd-Cr(VI)-Cu-Hg-Ni-Pb-Se-Zn)	3	$\frac{C_1}{L_1} + \frac{C_2}{L_2} + \frac{C_3}{L_3} + \ldots + \frac{C_n}{L_n}$ Assuming that the limit fixed for each single element should not be exceeded, the sum of the relationships between the concentration of each single element present (C) and the relative concentration limit (L) should not exceed 3. The limit refers to the element in solution as an ion in complex form and in suspension.
11	Aluminium (mg/l) as Al	1	The limit refers to the element in solution as an ion in complex form and in suspension after sedimentation of 2 hours.
12	Arsenic (mg/l) as As	0.5	The limit refers to the element in solution as an ion in complex form and in suspension.
13	Barium (mg/l) as Ba	20	The limit refers to the element in solution as an ion in complex form and in suspension after sedimentation of 2 hours.
14	Boron (mg/l) as B	2	The limit refers to the element in solution as an ion in complex form and in suspension after sedimentation of 2 hours.
15	Cadmium (mg/l) as Cd	0.02	The limit refers to the element in solution as an ion in complex form and in suspension.
16	Chromium (III) (mg/l) as Cr	2	The limit refers to the element in solution as an ion in complex form and in suspension after sedimentation of 2 hours.
17	Chromium (IV) (mg/l) as Cr	0.2	The limit refers to the element in solution as an ion in complex form and in suspension.
18	Iron (mg/l) as Fe	2	The limit refers to the element in solution as an ion in complex form and in suspension after sedimentation of 2 hours.
19	Manganese (mg/l) as Mn	2	The limit refers to the element in solution as an ion in complex form and in suspension after sedimentation of 2 hours.
20	Mercury (mg/l) as Hg	0.005	The limit refers to the element in solution as an ion in complex form and in suspension.
21	Nickel (mg/l) as Ni	2	The limit refers to the element in solution as an ion in complex form and in suspension.
22	Lead (mg/l) as Pb	0.2	The limit refers to the element in solution as an ion in complex form and in suspension.
23	Copper (mg/l) as Cu	0.1	The limit refers to the element in solution as an ion in complex form and in suspension.
24	Selenium (mg/l) as Se	0.03	The limit refers to the element in solution as an ion in complex form and in suspension.
25	Tin (mg/l) as Sn	10	The limit refers to the element in solution as an ion in complex form and in suspension after sedimentation of 2 hours.

INLAND WATERS AND DISCHARGES INTO PUBLIC SEWERS

Table A *continued*

No.	Parameter	Concentration	Notes
26	Zinc (mg/l) as Zn	0.5	The limit refers to the element in solution as an ion in complex form and in suspension.
27	Cyanide (mg/l) as CN	0.5	
28	Active chlorine (mg/l) as Cl_2	0.2	
29	Sulphides (mg/l) as H_2S	1	
30	Sulphites (mg/l) as SO_3	1	
31	Sulphates (mg/l) as SO_4	1000	Does not apply to discharges into the sea.
32	Chlorides (mg/l) as Cl	1200	Does not apply to discharges into the sea.
33	Fluorides (mg/l) as F	6	
34	Total phosphorus (mg/l) as P	10	The limit is reduced to 0.5 in the case of direct discharges into lakes or coastal waters within 10 km of the coastline.
35	Total ammonia (mg/l) as NH_4^+	15	For direct and indirect discharges into lakes within 10 km of the coastline all nitrogen (organic + ammonia + nitrous + nitric) must not exceed 10 mg N per litre.
36	Nitrous nitrogen (mg/l) as N	0.6	
37	Nitric nitrogen (mg/l) as N	20	
38	Animal and vegetable fats and oils (mg/l)	20	
39	Mineral oils (mg/l)	5	
40	Total phenols (mg/l) as C_6H_5OH	0.5	
41	Aldehydes (mg/l) as H-CHO	1	
42	Aromatic organic solvents (mg/l)	0.2	
43	Nitrous organic solvents (mg/l)	0.1	
44	Chlorine solvents (mg/l)	1	
45	Surface active substances	2	
46	Chlorine pesticides (mg/l)	0.05	
47	Phosphorus pesticides (mg/l)	0.1	
48	Toxic test		The sample diluted 1 : 1 with standard water must allow, in aerated condition, the survival of at least 50% of the animals used for the test, for a period of 24 hours at a temperature of 15 °C. The species used for the test should be *Salmo Gairdnerii* Rich.

EMISSION AND CONCENTRATION LIMITS

Table A *continued*

No.	Parameter	Concentration	Notes
49	Total Coli (MPN/100 ml)	20 000	The limits apply when, at the discretion of the controlling authorities, concommitant uses of the receiving water source require it.
50	Faecal Coli (MPN/100 ml)	12 000	
51	Faecal streptococchi (MPN/100 ml)	2 000	

The analytical determinations should be carried out on an average sample taken during a period of at least 3 hours. The methodical analyses of samples to be used in the determination of the parameters are those described in the volumes on 'Analytical Methods for Water' published by the Water Research Institute (CNR), Rome, and their subsequent updating.

For effluent from existing manufacturing installations discharging into surface watercourses there is a shorter time allowed for conformity with the permitted limits: within 3 years from the date of entry into force of Law 319 of 1976 the permitted limits laid down in Table C in the law must be adhered to (but this period has been extended to 1 March 1980 by article 1 of Law 650 of 1979).

Table C

No.	Parameter	Concentration	Notes
1	pH	5.5–9.5	The pH value of the receiving watercourse must be between 6.5 and 8.5 in a 50 m range from the discharge.
2	Temperature (°C)		For watercourses the maximum variation between the average temperature of any section of the watercourse above and below the discharge point must not exceed 3 °C. On at least half of any section below the discharge point the variation must not exceed 1 °C. For lakes the temperature of the discharge must not exceed 30 °C and the increase in temperature of the receiving watercourse must in no case exceed 3 °C at more than 50 m from the discharge point. For canals the maximum average temperature of the canal water at any point below the point of discharge must not exceed 35 °C. This is subject to the approval of the authorities in charge of managing the canal. For the sea the temperature of the discharge must not exceed 35 °C and the increase in temperature of the receiving source must in no case exceed 3 °C at a distance of more than 1000 m from the point of discharge.

INLAND WATERS AND DISCHARGES INTO PUBLIC SEWERS

Table C *continued*

No.	Parameter	Concentration	Notes
3	Colour		Not perceptible after dilution 1 : 40 over a width of 10 cm.
4	Odour		Should not cause inconvenience or nuisance of any kind.
5	Coarse materials	Absent	'Coarse materials' refers to objects of any nature with a linear dimension greater than 1 cm.
6	Sedimentary materials (ml/l)	2	Sedimentary materials are measured in an Imhoff cone after 2 hours.
7	Total materials in suspension (mg/l)	Not more than 40% of value above purification plant[a]	'Total materials in suspension', regardless of their nature, refers to materials of a size which does not allow them to pass through a filter membrane of 0.45 µm porosity.
8	BOD_5 (mg/l)	Not more than 70% of value above purification plant[b]	
9	COD (mg/l)	Not more than 70% of value above purification plant[c]	The COD should be determined with boiling bichromate of potassium after 2 hours.
10	Total toxic metals and non-metals (As-Cd-Cr(VI)-Cu-Hg-Ni-Pb-Se-Zn)	3[d]	$\dfrac{C_1}{L_1} + \dfrac{C_2}{L_2} + \dfrac{C_3}{L_3} + \ldots + \dfrac{C_n}{L_n}$
11	Aluminium (mg/l) as Al	2	The limit refers to the element in solution as an ion in complex form and in suspension after sedimentation for 2 hours.
12	Arsenic (mg/l) as As	0.5	The limit refers to the element in solution as an ion in complex form and in suspension.
13	Boron (mg/l) as B	4	The limit refers to the element in solution as an ion in complex form and in suspension after sedimentation for 2 hours.
14	Cadmium (mg/l) as Cd	0.02	The limit refers to the element in solution as an ion in complex form and in suspension.
15	Chromium (III) (mg/l) as Cr	4	The limit refers to the element in solution as an ion in complex form and in suspension after sedimentation for 2 hours.
16	Chromium (IV) (mg/l) as Cr	0.2	The limit refers to the element in solution as an ion in complex form and in suspension.
17	Iron (mg/l) as Fe	4	The limit refers to the element in solution as an ion in complex form and in suspension after sedimentation for 2 hours.
18	Manganese (mg/l) as Mn	4	The limit refers to the element in solution as an ion in complex form and in suspension after sedimentation for 2 hours.
19	Mercury (mg/l) as Hg	0.005	The limit refers to the element in solution as an ion in complex form and in suspension.

Table C *continued*

No.	Parameter	Concentration	Notes
20	Nickel (mg/l) as Ni	4	The limit refers to the element in solution as an ion in complex form and in suspension.
21	Lead (mg/l) as Pb	0.3	The limit refers to the element in solution as an ion in complex form and in suspension.
22	Copper (mg/l) as Cu	0.4	The limit refers to the element in solution as an ion in complex form and in suspension.
23	Selenium (mg/l) as Se	0.03	The limit refers to the element in solution as an ion in complex form and in suspension.
24	Zinc (mg/l) as Zn	1	The limit refers to the element in solution as an ion in complex form and in suspension.
25	Total cyanides (mg/l) as CN	1	
26	Active chlorine (mg/l) as Cl_2	0.3	
27	Sulphides (mg/l) as H_2S	2	
28	Sulphites (mg/l) as $SO_3^=$	2	
29	Sulphates (mg/l) as $SO_4^=$	1000	Does not apply to discharges into the sea.
30	Chlorides (mg/l) as Cl	1200	Does not apply to discharges into the sea.
31	Fluorides (mg/l) as F	12	
32	Total phosphorus (mg/l) as P	10	The limit is reduced to 0.5 in the case of discharges directed into lakes within 10 km of the coastline.
33	Total ammonia (mg/l) as NH_4^+	30	For direct and indirect discharges into lakes within 10 km of the coastline all nitrogen (organic + ammonia + nitrous + nitric) must not exceed 10 mg N per litre.
34	Nitrous nitrogen (mg/l) as N	0.6	
35	Nitric nitrogen (mg/l) as N	30	
36	Animal and vegetable fats and oils (mg/l)	40	
37	Mineral oils (mg/l)	10	
38	Total phenols (mg/l) as C_6H_5OH	1	
39	Aldehydes (mg/l) as H-CHO	2	
40	Aromatic organic solvents (mg/l)	0.4	
41	Nitrous organic solvents (mg/l)	0.2	
42	Chlorine solvents (mg/l)	2	
43	Surface active substances	4	
44	Chlorine pesticides (mg/l)	0.05	
45	Phosphorus pesticides (mg/l)	0.1	

Table C *continued*

No.	Parameter	Concentration	Notes
46	Toxic test		The sample diluted 1 : 1 with standard water in an aerated condition must allow the survival of at least 50% of the animals used for the test for a period of 24 hours at a temperature of 20 °C. The species used for the test should be *Carassius auratus*.
47	Total Coli (MPN/100 ml)	20 000	The limit applies at the discretion of the controlling authority when the concommitant uses of the receiving water source require it.
48	Faecal Coli (MPN/100 ml)	12 000	
49	Faecal Streptococchi (MPN/100 ml)	2 000	

[a] Minimum imposable limit: that of Table A; maximum limit: 200.
[b] Minimum imposable limit: that of Table A; maximum limit: 250.
[c] Minimum imposable limit: that of Table A; maximum limit: 500.
[d] Given that the limit for each single element must not be exceeded, the total of the concentrations in which each single element is present and the relative concentration limit should not exceed a value of 3.

The analytical determinations should be carried out on an average sample taken during a period of at least 3 hours. The methodical analyses of samples to be used in the determination of the parameters are those described in the volumes on 'Analytical Methods for Water' published by the Water Research Institute (CNR), Rome, and their subsequent updating.

With reference to civil installations discharging effluent into surface waters, article 14, clause 2 of Law 319 of 1976 should be examined, which places the control of effluent from civil installations not discharged into public sewers under the regional treatment plans. The objectives to be achieved by this control are those indicated in Table A. Nevertheless, while for existing civil installations it can be maintained that it is sufficient to wait for the regional treatment plans, where new installations are concerned it would seem more in line with the spirit of the law to require immediate conformity with the limits in Table A. The 'general technical regulations' on the nature and substance of disposal plants discharging on land or in the subsoil from civil installations of less than 50 rooms or 5000 m^3 (issued by the Committee of Ministers under article 2 of Law 319), which permit only certain forms of disposal, do in fact prohibit discharges into surface waters.

3.5.2 Discharges into public sewers

The effluent from new manufacturing installations discharging into public sewers is subject to the permitted limits contained in Table C, until the centralised purification plant begins operating (see 3.5.1). Once the centralised plant is functioning new permitted limits will be set by the authority in charge of the public sewerage and purification service. Pending the completion of the centralised purification plant, discharges from existing manufacturing installations must comply with the limits contained in Table C by 1 March 1980 (article 13 of Law 319 of 1976).

We have seen, however, that with the new regulations (Law 650 of 1979) the communes and consortia can dictate more rigorous requirements even before the centralised purification plants have begun operating (see 3.4). However, where discharges from existing manufacturing installations are subject to these more rigorous controls, the communes and consortia involved must complete their centralised purification plants within 18 months of regional approval of the requirements they dictated, and in any case not later than 31 December 1981.

The law does not impose limits on the effluent from civil installations discharging into public sewers but makes them subject to compliance with the regulations issued by the authority running the sewerage service (commune, intercommunal consortium).

Provisional authorisation under article 15, clause 8 of Law 319 of 1976 is subject to observance or progressive alignment with the tables attached to the law, or to the permitted limits set by the authorities managing the sewerage service.

3.6 POWERS OF THE PUBLIC AUTHORITIES

Law 319 of 1976 gives the public authorities vast powers, some of which, more precisely those creating instruments of inspection and control on the condition and conducting of discharges, have already been discussed under 3.3. It should be added here that the technical requirements for the control of purification installations set out by the Committee of Ministers have regulated the situation of stoppages on the part of purification installations due to breakdown or maintenance, which is not covered by the law: such an occurrence under the technical requirements must be immediately communicated to the controlling authorities. In this event, if the purification installation handles particularly toxic substances contained in list I annexed to the Directive of the Council of the EEC of 4 May 1976 (EEC/76/464), the authority can order the discharge to be suspended, thus compelling the manager to suspend the manufacturing cycle or to collect the effluent in suitable receptacles while waiting for the purification installation to resume operating.

Law 319 provides for a system of penal sanctions for breaches of the law requiring authorisation. An unauthorised discharge is punishable by imprisonment for a period of from 2 months to 2 years or by a fine of from 500,000 to 10,000,000 lire (this punishment applies not only to any owner of a discharge from a new installation, but also to anyone who has failed to apply for the authorisation or renewal required for existing discharges). Imprisonment is always applied if the discharge violates the permitted limits in the table attached to the law (see 3.5). Imprisonment for up to 2 years or a fine of up to 10,000,000 lire is also applicable for people who continue or carry out an authorised discharge without observing the precautions set out in the authorisation provision. Punishment by a fine of up to 5,000,000 lire is, instead, applied in the case of discharges being carried out before the authorisation applied for has been granted (assuming that it is granted, otherwise it comes into the category of an unauthorised discharge).

Of great importance is the provision contained in article 24 of Law 319 which states that the punishment can be conditionally suspended, subject to complete compliance with all the requirements contained in the judgement. However, it is not very clear what the content of such an injunction would be since it cannot consist of the violated provisions or exceeded limits with which the offender must in any case comply. In this context, the law requires the judge to request suitable indications from the administrative authority, which leads one to suppose that the injunction contained in the judgement is intended to result in compliance with those particular technical requirements necessary to re-establish the ecological balance of the polluted water sources.

Law 650 of 1979 has introduced into Law 319 of 1976 an article 23 *bis*, which provides for a fine of 100,000 to 1,000,000 lire for people who do not ensure that equipment to measure the quantity of water extracted is installed and functions correctly, or who do not comply with the requirements of the authority concerning the installation of such equipment.

In addition, there are offences covered by the Penal Code and used to control acts of water pollution in cases where an act occurs which in the abstract can be considered to fall into the category in question; this arose to control phenomena very different from pollution caused by civil or productive installations (see following list):

The Penal Code

Article 439: Poisoning of water or foodstuffs. Whoever poisons water or substances destined for foods before they reach the consumer or are distributed for consumption shall be punished by imprisonment for not less than 15 years. If the above act leads to the death of anyone, life imprisonment shall apply; and in the case of the death of more than one person, the death penalty shall apply (abolished).

Article 440: Adulteration and counterfeiting of foodstuffs. Whoever corrupts or adulterates water or substances destined for foods before they reach the consumer or are distributed for consumption, making them dangerous to public health, shall be punished by imprisonment of from 3 to 10 years. The same punishment shall apply to persons counterfeiting, in a way which is dangerous to public health, foodstuffs destined for commercial use. The punishment shall be increased if medical substances are adulterated or counterfeited.

Article 452: Crimes against public health. Whoever commits, by negligence, some of the acts under articles 438 and 439 shall be punished:

(1) by imprisonment of from 3 to 12 years in cases for which the death penalty (abolished) applies;

(2) by imprisonment of from 1 to 5 years in cases for which life imprisonment applies;

(3) by imprisonment of from 6 months to 3 years for cases punishable by imprisonment under article 439.

When any of the acts covered by articles 440, 441, 442, 443, 444 and 445 are committed by negligence, the penalty established under the above acts shall be applied, reduced from one-third to one-sixth.

Article 674: Dangerous throwing of objects. Whoever throws or pours objects liable to offend, soil or cause a nuisance to anyone in a place of public transit or in a private place used by others, or even, in cases not permitted by the law, causes emissions of gas, vapours or smoke liable to cause such effects, shall be punished by imprisonment of up to 1 month or a fine of up to 80,000 lire.

Article 675: Dangerous placing of objects. Whoever, without suitable precautions, places or suspends objects which, falling in a private place which is for common use or for the use of others, can offend, soil or cause a nuisance shall be punished with a fine of up to 40,000 lire.

Article 26 of Law 319 of 1976 expressly excludes 'specific and motivated or integrative intervention on the part of the competent health authorities' as far as drinking of water, mussel culture, bathing, and the protection of public health are concerned. This applies not only to local health authorities but to the whole of the health organisation; nevertheless, the sphere of regional intervention assumes particular importance through its organs, which hinge on the organisational structure of the region (particularly the Provincial Medical Officer).

3.7 RIGHTS OF THE INDIVIDUAL

See 1.4.

3.8 LIMITS AND SCOPE OF COMPLEMENTARY REGULATIONS

Law 319 of 1976 has rationalised the control of a sector which had previously been subject to numerous different regulations by repealing all previous provisions which either directly or indirectly controlled the question of discharges.

Nevertheless, we have seen that part of the preceding legislation has remained in force (Law 319 cites the penal regulations for the protection of life and personal and public safety and excludes the powers of intervention of the health authorities) and partly performs an integrative role useful for the more general purpose of the protection of the aquatic environment with regard to the control of discharges.

The laws under the TU of the inland fishing water laws and the TU of the health laws which were at one time in use for the prevention or restriction of acts of pollution no longer appear to apply since the entry into force of Law 319 of 1976. This also appears to be true for atypical discharges which do not arise from civil or manufacturing installations, for which it seems preferable to extend the scope of Law 319 by appropriate use of instruments such as the technical provision issued by the Committee of Ministers, rather than make use of the previous provisions. Discharges arising from ships, for example, are not controlled by Law 319 of 1976, nor are they covered in other legislation, apart from Law 59 Prov. Trento of 1973. For these and other problems the flexible structure of the law assumes a vital importance in relation to its response to protection requirements; among other things it allows the full and correct exercising of regional powers in, for example, matters of navigation on inland waters.

The regional laws on water pollution issued before Law 319 of 1976 remain in force if they are compatible (above all with regard to the way in which the various functions are distributed under Law 319). Article 1 of Law 319 also excludes the provisions under DPR 185 of 13 February 1964 (see Chapter 7 on Nuclear Energy).

An important series of regulations on the subject of water resource utilisation is constituted by the TU on water and electricity plants (RDL

1775 of 11 December 1933), containing provisions at times very similar to those of Law 319 of 1976. In particular, it states that applications for permission for concessions and new uses of public waters should be presented to the Civil Engineers accompanied by outline plans for the abstraction equipment, conduct, use and discharge of the water; that the regulator of the concession should determine the methods and quantities abstracted, the conduct, use and discharge of the water, with safeguards for public hygiene; that users must measure the water abstracted and discharged by means of instruments approved by the Italian Hydrographic Service. With the transfer of the duties of the office of the Civil Engineers to the regions a comprehensive function for the protection of water resources could be achieved and strengthened as a result of the powers granted to the regions under Law 319 of 1976 and TU 1775 of 1933. The problem, however, is one of coordination between the functions relating to the permission for derivation and new uses of public waters (which fall within the sphere of regional powers through the offices of the Civil Engineers) and the functions relating to the authorisation for the discharge and water quality control (which, under Law 319 of 1976, is the responsibility of local bodies). In this respect legislative intervention is required to complete the task undertaken in the project which resulted in the partial and restricted achievements of Law 319 of 1976: to reorganise the administrative functions along coordinated, decentralised lines, while re-examining the question of concessions and uses of public waters which today is controlled by the TU of the water laws of 1933.

Law 650 of 1979 has achieved a partial renewal of the administrative structures, aided by the simultaneous entry into force (1 January 1980) of the health reform law (Law 833 of 23 December 1978). Furthermore, powers are also given to the Ministry of Public Works and CIPE, particularly in respect to finance and investments which were shown to be weak points in the previous provisions (allocations of the State budget have already been fixed for 1980, 1981 and 1982; loans to local bodies are also provided by the Deposit and Loan Bank and decisions concerning minimum quotas and regional distribution are taken by CIPE).

In addition, the Government is required to refer to Parliament concerning the implementation of the law with a report annexed to the expenditure forecast of the Ministry of Public Works.

With regard to the acceptance of international agreements on the subject, Law 527 of 24 July 1978 should be noted. This law ratifies and renders executive the convention between Italy and Switzerland concerning the protection of Italo-Swiss waters from pollution, and was signed in Rome on 20 April 1972.

4
Coastal Waters and Open Seas

Legislation for the prevention and repression of marine pollution is subject to the peculiar characteristics of this type of pollution: polluting activities carried out in one country are liable to represent a risk to the environmental stability of various other countries bordering on the same sea, particularly in geographical environments such as Italy which is situated on an enclosed sea constituting an important water basin for the life of all the Mediterranean countries. It is therefore natural that, at an international level, controls should be adopted by means of international treaties or agreements to cover these supernational interests in marine pollution. On the other hand, the sea comes into the category of protected water sources on a par with international waters under the water protection legislation and control of discharges; however, for discharges into the sea certain exceptions have been introduced, mainly concerning the distribution of the different areas of responsibility among the public authorities (see 4.1.2).

4.1 COASTAL WATERS

This expression can be understood to mean the waters belonging to the public seaboard plus the territorial waters (roadways, ports, lagoons) and in any case those directly influenced by activities being carried out on the coast.

4.1.1 Applicable legislation

The protection of coastal waters in Italy is carried out through three basic regulatory instruments:

(i) The Ordinary Law provisions contained in the Navigational Code and aimed at hindering or repressing acts (rather than pollution in the technical sense) involving fouling or depositing of objects into the water (see articles 71 and 76 of the Navigational Code). These provisions are not very effective, both because they take into consideration occasional causes of water deterioration, often not particularly dangerous, and also because they only apply within the area of a port. Article 71 of the Navigational Code states that 'in ports it is forbidden to throw materials of any kind' into the water, leaving any extension of this prohibition to other areas to the discretion of the Head of the Maritime Department. Article 76 of the Code refers to industrial installations or storage depots sited on wharves, quays or navigable canals, and obliges the managers of such installations or depots to obviate the fouling of waters caused by these installations or depots, in accordance with the orders of the Head of the Maritime Department. The provisions contained in the Navigational Code are integrated and made more specific in the area of maritime navigation by the implementing regulation (DPR 328 of 15 February 1952). See article 77 (on-board wastes: prohibition to throw such wastes into the sea at a distance less than that specified by the harbourmaster) and article 82 (port cleaning is to be carried out by whoever undertakes loading or unloading operations and is extended to cover the stretch of water beyond the quayside). The provisions under the Navigational Code are also of significance with regard to the management, protection and accessibility of the public seaboard (article 28, ff.; 62, ff. Navigational Code; 59 ff. of the Regulation).

(ii) The provisions relating to sea fishing consisting of Law 963 of 14 July 1965; implementing regulation DPR 1639 of 2 October 1968; and amendments to the latter contained in DPR 1057 of 9 June 1976. Law 963 of 1965 contains some regulations for the protection of biological resources directly relating to those activities which are potential sources of marine pollution; in particular, article 15 concerns certain types of offence and especially crimes consisting of acts of pollution (see 4.1.5). Under article 14 this law refers the question of authorisation for the discharging of wastes into the sea to the implementing regulation, DPR 1639 of 1968, which remained in force as the only regulatory source concerning discharges into the sea until 1976. The legislation on fishing is generally a useful instrument and can also be useful for the protection of the coastal environment: in addition, an important aspect of this protection such as protection of the marine biological heritage, in spite of the fact that it is only a partial protection of ecological requirements, is that protection is carried out primarily through restrictions and prohibitions concerning fishing (e.g. the creation of a protected biological reserve in the strip of sea facing Portoferraio: DM of 10 August 1971).

(iii) The provisions specifically controlling discharges into the sea or coastal activities directly or indirectly polluting marine waters also apply to discharges into the sea (see Law 319 of 1976) (Chapter 3) since it controls 'discharges of any type, public and private, direct and indirect, into all surface and ground waters, inland and marine, public and private, as well as into sewers, land and subsoil', as modified and integrated by Law 650 of 1979 (see Chapter 3). However, the control of discharges into the sea largely differs from the control of discharges into other water sources, as will be shown in 4.1.3. Before Law 319 of 1976 was passed, the short Law 126 of 16 April 1976 was also concerned with discharges into the sea, trying to organise the subject until an all-encompassing law on the protection of waters is passed. In the same period there was another law concerned with the protection of the sea in the vicinity of the coast. Law 203 of 8 April 1976. This law controls authorisations for the planning, construction and management of installations for the collection and treatment of oily sludges, ballast and rinsing waters from ships and tankers. These installations, which were required by the International Convention for the Prevention of Pollution from Ships (London 1973), had in fact been recommended several years previously in a circular from the Minister for the Mercantile Marine on the basis of previous London Conventions (therefore in an administrative act without an imperative effect such as that of a normative act).

On the subject of international conventions relating to water protection and protection of coastal waters from oil pollution which constitutes another source of specific control, the situation is as follows: the International Convention signed in London on 12 May 1954 (International Convention for the Prevention of Pollution of the Sea by Oil) was put into effect by Law 238 of 23 February 1961; subsequent amendments to that convention, decided in London on 11 April 1962, were put into effect by Law 94 of 14 January 1970. The last adjustments were put into effect with Law 341 of 5 June 1974 and Law 875 of 19 December 1975. The Convention also signed in London on 29 December 1972 by Italy to which article 11 of Law 319 of 1976 refers has not yet been rendered executive. The other London Convention of 1973 was ratified and implemented by Law 662 of 29 September 1980. With regard to coastal protection, the agreement between Italy and Yugoslavia which came into force on 20 April 1977 rendered executive by DPR 992 of 29 May 1976 is also of significance. This agreement provides for a joint commission to be set up to protect the Adriatic Sea and coastal areas, with powers to examine and study pollution problems and to make proposals to the Governments concerning research projects and useful or necessary provisions; recently, Law 405 of 29 July 1981 provided for the financing of oceanographic research and studies to be carried out in implementation of the agreement. As far as the Tyrrhenian coast is concerned, Italy is also a party to the agreement between Italy, France and Monaco on the Protection of Mediterranean Waters which came into effect on 1 March 1981. In addition, it should be remembered that after Law 319 of 1976 was passed it was expressly stated in Law 690 of 1976 that regulations under Law 171 of 16 April 1973 (Law for Venice) and DPR 962 of 20 September 1973 (also concerned with safeguarding the environment of the Venetian lagoon) are still valid. Although it is not mentioned in Law 690 of 1976, it should be considered that Law 366 of 5 March 1963 on the same subject is also still valid, even after the new water protection legislation was issued. Law 650 of 1979 has also modified Law 171 of 1973. Finally, the provisions of the Decree from the Minister for the Mercantile Marine concerning nautical anti-pollution methods should also be noted. The latest provision of this kind is the DM of 6 September 1977 modifying the provisions previously issued under DM of 28 February 1975. See 2.2.3.1 regarding the implementation of EEC Directive EEC/439/75.

4.1.2 Competent authorities

The protection of marine waters, including coastal waters, formed a sector of State responsibility in which the activities of the regions and the other local bodies were of only indirect importance.

Even Law 319 of 1976 which, as we have seen in Chapter 3, grants important powers to the regions and assigns to the provinces, communes and consortia the responsibility for authorising discharges and controlling polluting activities under the law, introduces considerable exceptions to the system by restricting the relevant authority to State organs. Article 11 of Law 319 of 1976 in its original form stated that a discharge of waste from industrial processes or from public services or installations of any type directly into the sea must be authorised by the Head of the Maritime Department, thus retaining the competence provided for other similar authorisations under DPR 1639 of 1968 (implementing regulation of the law on fishing in maritime waters: see 4.1.1(ii)). The Committee of Ministers, in consultation with the Minister for Foreign Affairs and the regions involved, was responsible for granting authorisation for discharges into the open sea (this was pending the execution of the London Convention of December 1972 and a comprehensive international protection policy for the Mediterranean). The competence of the State and consequent exception to the regime provided for by the law for discharges which do not go into the sea has been justified, according to some experts, by the need for direct State management of State-owned property, including the seaboard and territorial waters (lagoons, coastal strip, ports etc.). A further justification is put forward: that the protection of coastal waters includes interests not restricted to the regions, but this is also valid for river estuaries, i.e. the Po, without this having prevented penetrating powers being given to the regions. But another justification appears to be decisive and this is the necessity for the State to adhere to certain rules relating to international relations and consequently the need to assume direct responsibility for activities involving the interests of other States bordering on the same sea, and also to control such activities in a direct manner, preserving the authority to achieve, even by coercion, certain results in accordance with international obligations.

This legislation has been radically altered by Law 650 of 1979 which has modified the text of article 11 of Law 319 of 1976, attributing the competence for authorising discharges directly into the sea 'to the authority designated by the territorially responsible region'; at the same time, the powers of the maritime authority have been preserved regarding control of the use of the seaboard and territorial waters, and navigation. In this way the view that protection of the sea and the coasts is an activity which transcends the interests of single regions has been surpassed, except for the need to take into account the international obligations of the State: the new text of article 11 states that the authority responsible for granting authorisations must inform the Ministry for the Mercantile Marine when each authorisation is granted for the purposes of notifying the relevant international organisations (see 4.2).

In addition, the local authority can make use of other effective instruments for the protection of coastal waters. The regions can pursue such protective objectives as a practical result of the undertaking of other powers delegated to them; there are regional laws concerned with town planning matters which prohibit installations of any kind on a given strip (of varying width) in the vicinity of the coast, and this acts as a contributory factor in the reduction of pollution; equally, there are regional laws concerning public works such as port installations which are concerned with providing instruments for environmental protection, and there are many other examples.

But above all, within the same specific legislation on discharges there is no distinction, for certain purposes and in defining certain powers of regions and other local bodies, already in the first text of the law, between discharges into the sea and discharges into other waters. Thus, article 8 of Law 319 of 1976 conferred on the regions preparation of the treatment plan with reference to all types of discharge. Article 13 of the same law specifies that the method and timetable for conforming to the limits set by the law for effluent from already existing manufacturing installations discharging into surface waters (including the sea) are established by the regional treatment plans; article 14 of the same law states that the control of effluent from civil installations discharged anywhere except into public sewers is established by the regional treatment plans, and so on. In addition, discharges into the sea deriving from public sewers constitute a type of effluent directly controlled by the authority which manages the service (communes, consortia, mountain communities): these authorities have the power, under articles 12, 13 and 14 of Law 319 of 1976, to issue provisions and regulatory specifications concerning the permitted limits for effluent from civil or manufacturing installations discharging into public sewers, and thus influencing the qualitative levels of waters discharging into the sea from public sewers and public purification plants.

The organisation of the State which has direct administrative responsibility in this matter is structured as follows:

(i) Ministry for the Mercantile Marine, which since 1947 (D. Leg. 376 of 31 March 1947) controls the Mercantile Marine in replacement of the Ministry for Communications (article 15, Navigational Code).

(ii) Maritime Director which is the peripheral office of the Minister for the Mercantile Marine, one in charge of each of the maritime divisions into which the Italian coast is divided.

(iii) Head of the Maritime Department which is a peripheral office in charge of a territorial area (Maritime Department) resulting from the subdivision of the maritime zones.

(iv) Head of the Maritime District, another peripheral office in charge of each district, resulting from a further territorial subdivision of the Departments.

In addition to these offices, in the major ports which are not maritime departmental or district offices, there are local port offices or shore representatives (article 16, Navigational Code).

All harbourmasters come under the Ministry for the Mercantile Marine with regard to their powers in matters of the seaboard and territorial waters: the harbourmaster's office supplies the personnel required for administrative functions relating to navigation and maritime traffic (article 18, Navigational Code).

4.1.3 Control of discharges

Law 319 of 1976 in its original form controlled 'the direct discharge into the sea of wastes from industrial processes, public services or from installations of any type'. Before that law had been passed, and with the exception of the controls under Law 126 of 1976 pending a comprehensive water protection law, DPR 1639 of 1968 controlled such discharges using a slightly different formulation: 'discharges into the sea of wastes from industrial processes or public services, however carried out'. In its new formulation of article 11 (introduced with Law 650 of 1979), Law 319 of 1976 refers quite simply to 'discharges directly into the sea'.

It was thought that the original wording of Law 319 of 1976 in its control of discharges was intended to include wastes from industrial processes not arising from manufacturing or civil installations in the sense in which such expressions are used by the law. That is to say, in its original formulation article 11 of Law 319 referred to two categories of waste discharged into the sea: waste from industrial processes and waste from public services or civil or manufacturing installations (the latter being separate from the first industrial category). It is possible to conceive of industrial activities not requiring installations and yet still being sources for the discharge of waste into the sea (fish processing on board ship, discharge of rinsing waters from petrol tankers etc.); if the original formulation of article 11 of Law 319 of 1976 specifically covered these two categories, it is difficult to understand the significance of the new formulation introduced by Law 650 of 1979. EEC Directive EEC/464/76 has so far not been put into effect.

4.1.3.1 AUTHORISATION FOR DISCHARGES

Authorisation for discharges into the sea is granted by the authority designated by the region (see 4.1.2) competent for the territory, which therefore assumes powers not only of authorisation but also of control (under article 9 of Law 319 of 1976). Under article 11 of Law 319 of 1976, the discharges for which the authority is competent are 'direct' discharges; it is evident, in fact, that if the discharge goes first into rivers, public sewers, the land or subsoil, it will come under the authority of the regions, communes or consortia as shown in Chapter 3. The new text of article 11 of Law 319 of 1976 (introduced by Law 650 of 1979) does not nominate the authority responsible for granting the authorisation, but leaves this designation to the relevant region to carry out. It can be considered that this designation should be carried out within the regional programming operation taking into account two of its characteristics: the reorganisation of the technical and administrative structures in charge of pollution control within the framework of the regional water treatment plan, and the reforms carried out within the health structure under Law 833 of 1978.

Law 319 of 1976 safeguards the powers of the maritime authority relating to the control and accessibility of the seaboard and territorial waters (article 11, clause 2). The Head of the Department ensures that the provisions governing utilisation of the seaboard and territorial waters are complied with (article 50, Navigational Code).

With regard to the structure of the authoritative provisions and procedures, see article 15 of Law 319 of 1976 which refers in general to all discharges under that water protection law and has already been discussed in Chapter 3. Thus, an authorisation or renewal has to be applied for from the relevant controlling authority (in this case, the authority designated by the region); the application should contain qualitative and quantitative characteristics of the discharge and the possible alternate destination permitted by law; a provisional authorisation for the discharge may be granted pending the final authorisation which is subject to compliance with the permitted limits under the law; the relevant authority must revoke the authorisation in the event of failure to comply with the limits laid down by the law; the cost of verifications, controls and on-the-spot investigations must be borne by the applicant.

Both the competence for the authorisation and the procedure are different if the discharge takes place in the open sea (article 11, clause 3 of Law 319 of 1976). In that case the third clause of article 11 of Law 319 of 1976 (as amended) is applicable; this article makes specific reference to discharges into the sea by ships or aircraft and confers responsibility for granting such authorisation on the Head of the Maritime Department (a peripheral office of the Ministry for the Mercantile Marine) responsible for that area (i.e. with administrative authority over the port from which the ship embarked or the nearest port (by aircraft) to the point of discharge). The authorisation granted must comply with the international conventions ratified by Italy, and with the directives established by the Interministerial Committee under Law 319 of 1976 (see 4.2.1).

The decision of the Committee of Ministers, Annex 5, 4 February 1977 formulated a detailed control of the discharge into territorial waters of residual sludge from purification plants for urban and industrial effluents and drinking water process discharges, as well as residual aqueous sludge from processing cycles. In addition to the information required by law, the application for an authorisation must contain an indication of the method of discharge and certain characteristics relating to the proposed zone into which the discharge will be made. The authorisation must contain a specification of the methods for carrying out all the necessary technical and scientific controls to ensure conformity with the regulations, conditions and restrictions imposed by the authority.

4.1.3.2 PERMITTED LIMITS FOR THE DISCHARGE

Article 11 of Law 319 of 1976 makes the granting of an authorisation conditional on conformity with the limits permitted under this law. Tables A and C annexed to the law and included in Chapter 3 also apply to discharges into the sea.

To summarise the position, the provisions concerning permitted limits for discharges are as follows:

(i) new manufacturing installations: obligation to conform to the limits shown in Table A (see Chapter 3) from commencement of operation;

(ii) existing manufacturing installations: obligation to conform provisionally to the limits shown in Table C (see Chapter 3) and subsequently, within 9 years from the entry into force of the law, to the limits shown in Table A, according to the methods and timetable of the regional treatment plans;

(iii) civil installations: obligation to conform to the provisions issued by the regions in the regional treatment plans, in the light of the directions established by the Interministerial Committee: these directives were established on 30 December 1980 by a Decree of the Interministerial Committee (G.U. 10 January 1981).

Progressive adherence to the limits in Table A is also required by the provisional authorisation for discharges which is granted pending full conformity with the limits under the law.

For discharges of residual sludge from treatment or purification plants, or from processing cycles, the decision of the Committee of Ministers of 4 February 1977, Annex 5, states that Table A and Table C must be observed in their deadline application 'except with regard to the content (in the sludge itself) of solid matter'. In this respect the requirements and limits defined by the Maritime Department in their authorising capacity have up to now been observed (it should be remembered that initially article 11 of Law 319 of 1976 safeguarded all the delegated powers of the Head of the Maritime Department regarding the protection and accessibility of the seaboard and territorial waters). In addition, it must be verified that the toxic substances contained in the sludge do not exceed, either in total or for single elements, the limits under nos. 10, 12, 15, 17, 20, 21, 23, 24 and 26 of Tables A and C, and also that the discharge must not give rise to alterations of a chemical, physical or biological nature of the marine environment such as to change the basic ecological structure and quality, and/or quantity of biological production, endanger fishing and/or other quantitative potential, cause the spread of pathogenic micro-organisms, damage the aesthetic aspect and tourism possibilities or cause inconvenience to marine traffic.

Still on the subject of sludge, there is also a geographic restriction for discharges: they must not be carried out at river mouths or in bordering coastal waters, in lagoons, in bays or partially enclosed areas where there is little movement to circulate the water, or in exploitation or breeding areas for edible shellfish.

4.1.4 Coastal activities relating to discharges and coastal pollution from fuels

We have seen (4.1.3) that, in addition to discharges arising from installations of any type (civil or manufacturing), article 11 of Law 319 of 1976 controlled emissions discharged into the sea of waste from industrial processes not arising from installations. This refers to a series of discharging activities not covered by the water protection law of 1976 and over which control is not clear since the entry into force of the new text of article 11 (with Law 650 of 1979).

First to be considered are the activities of plants and depots installed on the shoreline, wharves and quayside, covered by article 76 of the Navigational Code. This article obliges the operator to take precautions to avoid any fouling of the water which might arise from the plants or depots, or from an occasional but not continuous waste discharging activity.

But above all, reference should be made to industrial operations not carried out in installations, i.e. the discharge at sea in the proximity of the coast of oily sludge, ballast waters or rinsing waters from petrol tankers. This matter is one subject of the London Conventions for the Prevention of Pollution of the Sea by Oil signed by countries belonging to the Intergovernmental Maritime Consultative Organisation (including Italy). The text in force today and put into effect in Italy results from the Convention of 1954, amended in April 1962 (implemented in Italy by Law 94 of 14 January 1970), in October 1969 (Law 341 of 15 June 1974) and in 1971 (Law 875 of 19 December 1975). Article III of the Convention forbids the ships to which it applies to discharge mixed hydrocarbons, unless certain conditions are satisfied, among which is the distance from the coast and compliance with certain restrictions concerning the quantity of discharges (instantaneous flow of 60 litres per mile) and quality (content of hydrocarbons less than 100 parts per million of mixture), etc. More precise and rigorous requirements are established for tankers. Article VII of the Convention imposes requirements on ships to avoid as far as possible escape of hydrocarbons into the bilges, and also states that the transportation of ballast waters in tanks used for fuel should be avoided if possible. Article VIII of the Convention requires the signatories to take appropriate measures to promote the creation of port installations with adequate facilities for receiving mixed hydrocarbon residues which ships of every type have to discharge; installations to avoid the dispersion in the sea of wastes and oily sludges at points where hydrocarbons are loaded; etc.

Law 203 of 8 April 1976 refers to installations under the London Convention of November 1973. These are collection and treatment plants for oily sludges, ballast water and rinsing waters from petrol tankers. Planning, construction and management of these installations is granted, as a concession by decree from the Minister for the Mercantile Marine, to companies with State participation which run the dry docks and repair workshops in the ports of Genoa, La Spezia, Leghorn, Naples, Palermo, Taranto, Venice and Triest. The law has provided for a period of 120 days from its entry into force for terms to be agreed between the Minister and the relevant company with State participation.

The conventions also indicate the methods and conditions for utilisation of the installations on the part of petrol tankers in transit or loading or unloading in Italian ports. Plans for the installations are presented by the concessionary companies to the office of the Civil Engineers responsible for the territory in question. Authorisation for these projects is granted by decree from the Minister for Public Works and the Minister for the Mercantile Marine, after consultation with the autonomous bodies or consortia of the ports involved and the responsible consultative organs of the Ministry for Public Works.

4.1.5 Powers of the public authorities and obligations on the part of individuals

According to the system under Law 319 of 1976, the authority which is competent for authorising the discharge (for coastal waters, the authority designated by the region) also has the power of controlling the way in which the discharges are carried out, in addition, of course, to the power to revoke the authorisation. Therefore, the authority has the power to carry out all the inspections considered necessary inside manufacturing installations, to verify the conditions which give rise to the formation of the discharges, as well as the power to request that particular effluents containing substances listed under point 10 of Tables A and C (see Chapter 3) undergo special treatment before joining the main discharge. There is also an obligation on the part of the person responsible to make the discharges available for inspection and not to dilute the discharges with water extracted for the sole purpose of meeting the permitted limits under the law.

The penal provisions under Law 319 of 1976 regarding unauthorised discharges or violations of requirements laid down in the authorisation are also applicable to the discharges in question, because of the authorisation procedure (see 3.6).

A series of provisions in the Navigational Code also appears to apply to the control of certain acts of pollution, even if only occasional or exterior and limited to the area of the port. Thus, in the event of damage to the operation or installations of the port, the Head of the Department ascertains the extent of the damage through the offices of the Civil Engineers and instructs the person responsible to carry out, within a specific period, the necessary repairs (or alternatively to pay the costs of the repairs ordered by the Head of the Department and carried out by him in cases of urgency or failure to carry out the repairs in the specified time) under article 75 of the Navigational Code. Article 76, which applies to operators of industrial installations or depots on the shoreline, quayside or canals, is even more important, since it authorises the Head of the Department to take the necessary steps, at the expense of the party responsible for failure, to comply with the obligation to take good care of the sea bed and to avoid fouling the water.

Article 15 of Law 963 of 1965 on fishing in maritime waters covers two types of criminal offence under which marine pollution can be repressed. According to a combination of article 15 (*d*) and article 25 of Law 963 of 1965, unless it is not a serious offence, damage to the biological resources of the sea by the use of explosives, electricity or toxic substances which paralyse, stun or kill the fish and other aquatic organisms is punishable by up to 2 years imprisonment or a fine of up to 1,000,000 lire. This, however, applies only to the act of damage and not to the danger of damage, and requires actual damage to biological resources; in addition, it only applies to acts of intent.

The other form of offence under article 15 (*e*), calling for the same punishment, is of greater interest. In this case, the act of discharging, directly or indirectly, or of pouring polluting substances into the water is punishable. The provision defines such substances as extraneous substances, or substances which are part of the normal composition of natural waters, which constitute a direct injury to the aquatic wildlife or which cause such chemical or physical alterations to the environment as to influence negatively the life of the aquatic organisms. It must be believed that the offence in question is a criminalisation of the risk of pollution. The reference by the law to direct injury to the aquatic wildlife or to the chemical and physical alteration of the environment is aimed solely at identifying the polluting substances through their external effects. In this case, too, the offence only applies to acts of criminal intent, which leads to difficulties in the application of this regulation to the subject of discharges. If, in principle, the authorisation for a discharge does not exempt the discharger from penal or civil responsibility, the granting of an authorisation can induce the judge to consider the discharge of polluting substances into the sea a culpable act rather than an act of criminal intent. However, these difficulties were overcome when this provision was applied to the discharge into the sea of industrial process residues from Scarlino. In fact, this renowned case represents one of the experiences most symptomatic of the management difficulties on the part of an administration which is still not sufficiently cohesive and is unprepared to confront the problems of coastal water protection. This case also served to create one of the major stimuli for the creation of a comprehensive control of discharges and for a reorganisation (still not achieved) of the administrative structure.

4.1.6 Rights of the individual

See 1.4.

4.1.7 Quality standards for coastal waters

The provisions referred to above should be compared with the objectives and requirements resulting from the EEC directives on the subject.

A first coordination problem concerns the system introduced with Law 319 of 1976, in the light of EEC Directive 464 of 4 May 1976 ('Pollution caused by certain dangerous substances released into the aquatic environment of the Community'). This also applies not only to discharges into coastal waters but also to inland surface waters, inland coastal waters (e.g. lagoons) and groundwaters (article 1, Directive EEC/76/464).

The directive contains two lists of substances: the first list is of particularly harmful substances, the values of which are fixed in relation to their toxicity, persistence and bioaccumulation; the second list is of substances which have a harmful effect on the aquatic environment, but an effect which can, however, be restricted to a certain zone and are dependent on the characteristics and location of the receiving waters.

For the substances in the first list, it has been established that:

the discharge must be authorised beforehand;

the authorisation is granted for a limited period and is renewable, taking into account eventual alterations to the limit values;

those responsible for existing discharges must conform to the conditions established in the authorisation within the stipulated deadline.

Under article 6 of the Directive the Council shall set limits for the values which must not be exceeded in the discharges, and the quality objectives for the substances in the first list. Once the values have been fixed, a discrepancy can arise between these values and the tables contained in Law 319 of 1976: this eventuality is covered by article 3 of Law 319 of 1976 which leaves the revision of the tables to the Committee of Ministers (now the Interministerial Committee). Discrepancies can also arise in the nature of the authorisation required under EEC Directive 76/464 for every discharge and also required under Law 319 of 1976 as a 'verification that the discharge conforms with the limits of the plan' (CURTI GIALDINO) but at least from a formal viewpoint the discrepancies between the EEC directive and Law 319 do not appear significant.

Another EEC directive (8 December 1975, EEC/76/160) concerns the quality of bathing water and requires member States to make the quality of such waters conform to the limit values fixed on the basis of criteria and parameters established by the directive (article 3 and attached table) within 10 years from notification of the directive. The table establishes the parameters, values, frequency of sampling, methods of analysis and inspection. It is therefore a different regulatory perspective to that relating to discharges or other polluting activities since it takes into direct consideration the qualitative standards of the asset it is protecting (coastal and inland bathing waters). This directive is awaiting the promulgation of a law as delegated by Law 42 of 9 February 1982.

4.2 OPEN SEAS

4.2.1 Discharges in open seas of substances other than hydrocarbons

In this section we shall consider the discharge into the open seas arising from offshore installations (for discharges into coastal waters and the protection of the seaboard and coastal waters see 4.1.3), and from ships and aircraft (dumping), excluding pollution from hydrocarbons (for which see 4.2.3 and 4.2.4).

4.2.1.1 STATE REGULATIONS AND INTERNATIONAL CONVENTIONS

We have already mentioned (4.1.3) that discharges into the open seas are controlled by article 11 of Law 319 of 1965 (law for protecting waters from pollution), which, in clause 3, sets out particular competences and procedures for the authorisation and control of discharges.

The law states that the authorisation for the discharge is granted in conformity with the International Conventions in force and ratified by Italy (according to the directives established by the Interministerial Committee). See 2.2.3.1 regarding the implementation of EEC Directive EEC/ 439/75.

The international provisions should therefore be taken into consideration, all the more so because comprehensive international regulations for safeguarding the Mediterranean which could act as an international framework for the application of Law 319 of 1976 had already, at the time of the issuing of Law 319, become concrete in the Convention for the Protection of the Mediterranean Sea Against Pollution and in the protocols adopted in Barcelona on 16 February 1976, ratified in Italy by Law 30 of 25 January 1979. It should also be remembered that the Convention on the High Seas (adopted in Geneva in 1958) to which Italy is a party requires the signatory States to cooperate with the relevant international organs in carrying out measures for the prevention of marine pollution, resulting not only from any activity involving the use of radioactive materials (see 4.2.3.2) but also from other harmful agents.

The Convention of London of 1972 (to which Law 319 of 1976 referred before the amendment introduced by Law 650 of 1979) (i.e. the International Convention on Dumping of Wastes and other Matter at Sea), which was open for signature in London, Mexico City, Moscow and

Washington on 29 December 1972, has not yet been ratified by Italy which has, however, initiated the ratification procedure. This Convention establishes three categories of waste. The discharge of wastes included in the first category (mercury, cadmium and their compounds; plastics and persistent synthetic materials; most hydrocarbons and also heavily radioactive substances; etc.) is absolutely forbidden. Wastes included in the second category (containing arsenic, lead, zinc and their compounds; cyanide, fluoride, chrome, nickel, vanadium and other substances) can be discharged into the sea subject to rigorous controls and to the previous granting of a special permit. The third category of wastes includes substances which do not come into the first two categories: their discharge into the sea is subject to a general discharge permit.

The Convention requires that the responsible national authority granting the discharge permit, when examining the application, should take into account certain specific factors such as the possibility of synergistic effects and the eventual consequences of the discharges on other legitimate uses of the sea. In Annex III of the Convention the characteristics of the discharge, the location and method of the deposit etc. are specified.

Very similar requirements concerning the discharge of waste substances into the sea are contained in the Barcelona Convention of 16 February 1976. This Convention, after defining the geographical area corresponding to the Mediterranean Sea for the purposes of applying the agreement, and having made provisions concerning the stipulation of further bilateral or multilateral agreements, of a regional or sub-regional character (in line with the convention framework agreed at Barcelona, also in respect of the protocols already adopted and of those which can be added), requires the parties to carry out, individually or jointly, all the appropriate measures according to the requirements of the Convention to prevent and combat pollution of the Mediterranean and to protect the marine environment in this area. Such measures are also to be promoted within the international organs recognised by the parties to the Convention as being competent to deal with the matters in question. Within this framework the Convention provides for a series of sectors for intervention (pollution caused by discharges from ships and aircraft, pollution from shipping, from exploration or exploitation activities on the continental shelf and the sea bed, pollution from land-based sources etc.), as well as a series of initiatives of technical, scientific and juridical collaboration (formulation and adoption of appropriate procedures regarding responsibility and compensation for damage from marine pollution). The Convention then provides for a series of institutional mechanisms: the function of the Secretariat is delegated to UNEP (United Nations Environment Programme); institutes meetings of the parties in ordinary and extraordinary sessions to control the application of the agreements; provides for reports to the Organisation from the signatory countries; delineates arbitration procedures to settle controversies on the interpretation of the Convention itself or of its protocol; etc.

At this point we will only consider the discharge into the sea of damaging substances (other applications of the Convention will be discussed under the appropriate heading); article 8 of the Convention (regarding pollution from land-based sources) and article 5 (concerning the dumping or discharge of substances from ships or aircraft) must be considered.

Article 8 obliges the parties to carry out all the appropriate measures to prevent, abate and combat pollution of the Mediterranean Sea Area caused by discharges from rivers, coastal establishments or outfalls or emanating from any other land-based sources within their territories. Article 5 imposes similar requirements concerning dumping from ships or aircraft, but is completed by a protocol which is an integral part of the Convention. The protocol provides for three categories of wastes similar to the London Convention of December 1972 (referred to in the preamble to the protocol itself). The first category of wastes includes those containing substances listed in Annex I to the protocol; the discharge of these substances is absolutely forbidden. Annex II contains a list of substances which can be discharged into the sea subject to obtaining a special authorisation beforehand. The other substances (not included in the first two annexes) can be discharged into the sea if a general permit has been obtained previously from the authorities; Annex III establishes the factors to be taken into consideration when granting special or general permits. The system is similar to that of the London Convention of December 1972, even with regard to the lists of substances divided into three categories and the significant criteria for the decision regarding the discharge permit (Annex III: they are separated into factors relating to the properties and composition of discharged materials, to the characteristics of the discharge location and of the method of depositing, and to general considerations and conditions). The provisions described do not apply in the case of *force majeure.*

A comparison between this international regulation and that of Law 319 of 1976 gives rise to certain interesting considerations regarding the (remote) executiveness of the conventions described above in Italy.

The system provided for under article 11 of Law 319 of 1976 (granting by the Head of the Maritime Department of specific authorisation for each discharge into the open sea) conforms to the requirements of the Conventions signed by Italy. Control of the quantity, quality and method of discharge results from three references made: to the International Conventions, the directives of the Interministerial Committee, and the law (thus overcoming the previous system which made the Committee of Ministers competent according to the original regulation under article 11, old text, Law 319 of 1976, for granting the authorisation, and for determining the timing, the requirements and the restrictions to which the discharge is subject).

Article 11 of Law 319 of 1976, clause 3 as amended, takes into consideration discharges into the sea by ships or aircraft, whereas the previous text simply referred to 'discharges into the open sea'; it can therefore be maintained that some activities falling in the old system under the special controls over discharges into the open sea, are today controlled by the general regulations on discharges into the sea (see 4.1), whilst the law excludes other, more or less accidental, factors which do not constitute a discharge (immissions, spillages, jetsam etc.). These come under the general legislation in force before Law 319 of 1976; they can constitute an offence (see 4.1.5), they can come under the Navigational Code, etc.

With regard to the effectiveness of the provisions under Law 319 of 1976 it can be maintained that the obligation to obtain an authorisation for a discharge applies both for ships flying the Italian flag and for Italian citizens who carry out discharges into the sea by whatever means. In the protocol of the Barcelona Convention concerning dumping from ships or aircraft, the parties commit themselves to apply the agreed requirements to:

(i) ships and aircraft registered in their territory or flying their flag;

(ii) ships and aircraft loading in their territory wastes or other matter which is to be dumped at sea;

(iii) ships and aircraft believed to be engaged in dumping in areas under their jurisdiction in this matter.

A draft law was recently presented to Parliament on the initiative of the Government, containing 'provisions for the protection of the sea' (Senate Act 853, presented on 19 March 1980). The provisions it contains, which were laboriously arrived at by eleven ministers, will presumably not be approved before the end of 1982, but they should nevertheless be mentioned in this study since they represent an organic attempt at defining and implementing an effective policy for the protection of the marine environment. In fact, they cover the entire problem in an attempt at resolving the fragmentary nature of administrative roles and responsibilities and to insert the internal regulation in an organic and complete way into the context of the multitude of international provisions existing in the Mediterranean.

It was considered essential for the planning of intervention for the prevention of marine pollution to reaffirm the political responsibility of the Minister of the Mercantile Marine, backed by a suitable Interministerial Committee, entitled the National Committee for the Protection of the Sea. This Committee will be presided over by the Prime Minister, or as his delegate, by the Ministry of the Mercantile Marine, and will be made up of the Ministers of Foreign Affairs; the Interior; Defence; Health; Public Works; Industry, Commerce and Craft; the Mercantile Marine; Scientific Research; Environmental and Cultural Goods; and the Treasury. The Committee can include ministers responsible for individual matters under discussion.

The Committee will be assisted by an (advisory) council nominated by decree from the Minister of the Mercantile Marine and made up of:

three university lecturers in subjects relating to the marine environment;

the President of the Higher Council of Public Works or his delegate;

the President of the Higher Council of the Mercantile Marine or his delegate;

two representatives from the National Research Council;

one representative from the Higher Institute of Health;

one representative from the Marine Hydrographic Institute;

one representative from the regions appointed by the Inter-regional Consultative Committee under article 9 of Law 48 of 27 February 1967;

the Director of the Study and Experience Centre of the Fire Brigade;

one representative from the National Hydrocarbon Organisation (ENI);

three experts;

a magistrate from the Council of State;

a magistrate from the Court of Accounts;

a lawyer from General State Legal Profession.

The Committee's functions are as follows:

(i) to direct, promote and coordinate public and private activity in connection with the protection of the sea and the coast from pollution, with the exception of the control of discharges into the sea which are covered by Law 319 of 10 May 1976 and its subsequent amendments;

(ii) to elaborate and present proposals of a legislative and regulatory nature on the subject, making the lines of common action of the competent administrations agree with regard to operational intervention;

(iii) To prepare and coordinate study and research initiatives for the collection of data on the conditions of the marine environment.

The law increases the duties and responsibilities of the Minister of the Mercantile Marine who is charged with instituting a coastal monitoring service and intervention for the prevention and control of pollution at sea. With regard to accidents or imminent danger of pollution at sea, the maritime authority within whose jurisdiction the accident occurs must take all necessary measures to prevent, reduce or eliminate the effects of pollution of the sea water caused by hydrocarbon immissions or other harmful substances from any source liable to damage the marine environment, the coast or associated interests.

If the danger of pollution or act of pollution results in an emergency situation, the head of the maritime department will declare a local emergency; he will inform the Minister of the Mercantile Marine immediately, and will direct operations on the basis of the local plan for emergency intervention, leaving the powers of each administration in the execution of their duties unchanged. Finally, severe punishments have been introduced for offenders, in line with the international agreements on the subject.

The Government will be delegated to carry out the new duties envisaged in the text of the law. It will issue provisions concerning the revised area of responsibility of the Mercantile Marine and the structure of the central administration.

4.2.1.2 AUTHORISATIONS AND CONTROLS OVER DISCHARGES

The authorisation to discharge in the open sea is the competence of the Head of the Maritime Department of the relevant territory if the discharge is carried out by ships or aircraft. In the case of discharges in the open sea not arising from dumping activities from ships or aircraft, the general provisions for the control of discharges into the sea apply (authorisation from the authority designated by the region competent for the relevant territory).

As already mentioned, the Head of the Department responsible for granting the authorisation is the person in charge of the port from which the discharging ship sailed, or the person who administers the Department nearest to the point of discharge by the aircraft.

Before the amendment under Law 650 of 1979, Law 319 of 1976 stated that authorisation should be granted by the Committee of Ministers after consultation with the region involved. This consultation constituted a compromise solution to the conflict between State and region concerning discharges into the open sea. Before Law 319 of 1976 was passed this question was raised by the Region of Sardinia in the Constitutional Court which decided that the State was responsible for controlling the discharge of industrial residues into the open sea (sentence no. 203 of 1974). The justification adopted by the Court (and which gave rise to much criticism: GAJA) was based on the need of the State to exercise direct control over an activity which involved international commitments and for which it was directly responsible towards other States. In the original system, of Law 319 of 1976, a similar solution existed in the State's general responsibility for matters concerning discharges into the sea, both in coastal waters and the open sea (see 4.1.2). Today, on the contrary, the responsibility of the region has been extended to include direct discharges into the sea, leaving the State responsible for the authorisation and control of activities which concern the marine environment rather than safeguarding of the coast.

The Interministerial Committee defines the directives with which the authorisation must comply in order to conform with the International Conventions and the requirements of Law 319. This Committee exercises authority with very wide discretionary powers, and is aided, however, by technical backing supplied by the technical offices and organs of the Interministerial Committee. It should be remembered that the technical organ of the Committee is the Higher Council of Public Works and that the Committee makes use of the scientific and technical collaboration of the Higher Health Institute and of the laboratories of the Water Research Institute (IRSA) of the National Research Council (CNR).

The application is investigated by the Head of the Maritime Department, which must also inform the Ministry for the Mercantile Marine of each authorisation granted for the purposes of notifying the international bodies provided for by the conventions.

The general principle contained in article 9 of Law 319 of 1976 (by which the authority responsible for authorising the discharge and the controlling authority coincide) gives rise to various delegated powers, instrumental in the carrying out of the authorisation procedure, in favour of the Head of the Department: controls (sampling, inspections, analyses) required under Law 319 of 1976 are in this case undertaken to ensure that the requirements and restrictions imposed on the discharge by the authorisation provision are adhered to.

4.2.1.3 QUALITY AND QUANTITY OF DISCHARGES

Article 11, clause 3 of Law 319 of 1976 refers, in the new text, to the requirements of the International Conventions, to the directives of the Interministerial Committee, and to the provisions of the law, as distinct sources of regulations for discharges.

The International Conventions of London (1972) and Barcelona (1976) establish, as already mentioned, the substances which can be emitted, the conditions to which the discharge is subject and the criteria and characteristics of the discharge to be taken into account when assessing the application for authorisation: precise qualitative and quantitative limits for the discharge would therefore derive from the implementation of these Conventions.

Until then, the actual implementation of controls conventionally adopted at international level can be effected through spontaneous updating of the regulations by the Interministerial Committee when deciding upon the authorisation directives for the discharge.

It should, however, be noted that the use of internationally adopted regulations or regulations defined by the Interministerial Committee can not result in an exceptional lowering of the permitted limits, and worsening methods and conditions for the discharge specified in Law 319 of 1976; this is a result of the explicit provisions of the new text of article 11 of that law.

4.2.1.4 POWERS OF THE PUBLIC AUTHORITIES AND OBLIGATIONS ON THE PART OF THE DISCHARGER

The offences under article 15 of Law 963 of 1965 (on fishing in maritime waters) also apply to the subject of discharges in the open sea and those activities which do not come under Law 319 of 1976 and its relative sanctions (therefore occasional emissions, depositing, dumping etc.). In this case Italian penal law applies if the action, even partially, or the event either takes place in Italian territorial waters, or if the act is committed by Italian citizens, even outside territorial waters.

For activities constituting 'discharge' within the meaning of Law 319 of 1976, the penal regulations imposed by the law itself (articles 21–24) apply for cases of unauthorised discharge or violation of an authorisation (see 3.6).

One of the requirements of the Barcelona Convention of February 1976 on the control of pollution in the Mediterranean is that the parties to the Convention designate the competent authorities responsible for the control of pollution within the area of their national jurisdiction. In this respect the present situation in Italy is characterised by the presence of sufficiently effective methods of monitoring and repression of acts of pollution within the territorial waters, which have been the responsibility since 1973 of the Customs and Excise force (which is institutionally in charge of the surveillance of territorial waters during the course of its principal duties of customs and excise policing). The special air-survey service for protection of the coast and territorial waters of Italy, undertaken by the Customs and Excise force based on a project by the Ministry of the Environment is, however, almost exclusively concerned with pollution from hydrocarbons produced by ships or tankers discharging rinsing residues or hydrocarbon mixtures into the sea (see 4.2.3).

4.2.2 Pollution caused by the exploitation of the sea bed and offshore installations

4.2.2.1 STATE REGULATIONS AND INTERNATIONAL CONVENTIONS

The exploitation of recoverable resources from the sea bed is the object of provisions under Italian legislation, above all with regard to the exploration and exploitation of hydrocarbon deposits in territorial waters and the continental shelf. This matter is regulated by Law 613 of 21 July 1967, amended several times (DPR 1336 of 30 June 1973; Law 443 of 4 June 1973) but never for the purpose of taking into account the polluting effects of exploration, research and exploitation activities; neither is this aspect taken into consideration in the general context of the law. Nevertheless, the law contains some provisions which are useful for the regulation of environmental protection, and others which can be used by the responsible authorities to protect the sea.

First of all the law defines the continental shelf for its own regulatory purposes as the sea bed and its sub-soil adjacent to the territory of the Italian peninsula and islands and situated outside the territorial waters, up to a limit corresponding to 200 m depth or, beyond that limit, up to the point at which the depth of the water above still allows the natural resources of the area to be exploited. Article 1 of Law 613 of 1967 specifies that the delimitation of the external limit of the continental shelf shall be effected through agreements with States whose coasts face those of Italy (see the agreement between Italy and Yugoslavia concluded in Rome on 8 January 1968 and put into effect by DPR 830 of 22 May 1969).

Articles 2 and 4 of Law 613 of 1967 are of great significance, introducing aims and attributions which can be used by the maritime authorities as a means of protection from marine pollution. Article 2 reserves to the State the right to explore the continental shelf and to exploit its resources (thus confirming the basis of the permit system under the law for exploration, research and exploitation activities), establishes that such activities must be carried out in such a way as to avoid unjustified restrictions to navigational freedom, fishing, conservation of the biological resources of the sea, to other uses of the open sea according to international law as well as the conservation of the coast, beaches and ports. The regulation conforms (as do others referred to up to now) in its formulation to the text of the Geneva Convention of 1958 on the continental shelf, to which, however, Italy is not a party. Clause 5 of article 2 of Law 613 of 1967 makes the maritime authority competent for authorising or permitting the exploration of the continental shelf for ends other than those provided for in the law, or the exploitation of resources other than hydrocarbons and other mineral substances. Under article 4 of Law 613 of 1967 the maritime authority protects the rights of the State on the continental shelf and, in addition, by virtue of a regulation which can be used as an instrument for intervention by an authority with general competence for protection of the sea, ensures that authorised concessionaires are observing the obligations and restrictions imposed upon them by the Minister for the Mercantile Marine

The Minister for the Mercantile Marine in fact intervenes, as does the Minister for Industry, in the issuing of authorisation permits and concessions (granted jointly by the two ministers) with regard to the requirements of article 2 of Law 613 of 1967 for the protection of fishing, of the biological resources of the sea and the other uses of the open sea. According to the procedure for granting research permits, when the application is received, the Minister for Industry makes his decisions known to the Minister for the Mercantile Marine, to determine the obligations and restrictions to be imposed and on which the granting of the permit is conditional. These requirements are in force for exploration permits (article 9 of Law 613 of 1967), research (article 16) and exploitation concessions (article 27) as well as applications to increase the area of the concession (article 36). Article 40 of Law 613 of 1967 requires, in addition, approval by decree from the Minister for Industry, having heard from the Technical Committee for Hydrocarbons, of the type regulation for the permits or concessions in which are determined the particular conditions and methods for carrying out the exploration, research or exploitation activities (the type regulation was approved by DM of 29 September 1967). This can also be used as an instrument for imposing precise restrictions for protecting the sea. Nevertheless, as with the other regulations already discussed, the same argument applies: in the absence of precise legislative requirements (apart from the criteria established for the purpose by article 2 of Law 613 of 1967 mentioned above), everything is left to the goodwill and efficiency of the public administration, the discretionary powers of which are frequently of little use in confronting problems for which it is not prepared and which go beyond the traditional and normal protection of 'fishing' or of the 'biological resources of the sea', or even the 'coast, beaches, and coastal strips' etc. It is to be hoped that more effective controls will be adopted under the influence of international responsibility on the subject: the Barcelona Convention of 16 February 1976 requires signatory States to take all appropriate measures to prevent, abate and combat pollution in the Mediterranean area caused by the exploration and exploitation of the continental shelf, the sea bed and its sub-soil (article 7 of the Convention).

The Mediterranean signatory countries which subsequently agreed upon a plan of action to safeguard the Mediterranean, have also agreed to establish a protocol among the coastal States concerning preventive measures against pollution arising from the exploration and exploitation activities of the sea bed within the limits of their respective national jurisdiction. The discussion of this protocol should take place in 1983 (see 'Meeting of Experts on the Legal Aspects of the Pollution of the Mediterranean Resulting from Exploration and Exploitation of the Continental Shelf, the Sea Bed and its Sub-soil in the Mediterranean', Rome, 11–15 December 1978).

4.2.2.2 OBLIGATIONS ON THE PART OF OPERATORS

Among the obligations imposed on permit-holders and permitted concessionaires by Law 613 of 1976, in relation to the administrative activity for the prevention of pollution, there are a few which can constitute restrictions for the protection of the sea and which are fundamental to the protection of fishing, of marine biological resources, of the legitimate uses of the open sea and of the seaboard and territorial waters imposed by article 2 of the law.

In this way, the authorised research operator must observe the requirements laid down by the maritime authority with regard to matters under clauses 3 and 5 of article 2 of Law 613 of 1967 (article 22, no. 7 of the law) and an equivalent obligation is imposed on concessionaires of exploitation rights (article 30, no. 6).

Article 49 of Law 613 of 1967 is of importance, specifying that research and exploitation installations of the Italian continental shelf are subject to the laws of the State; Italian law therefore regulates any act which occurs on the continental shelf, regardless of the regulation valid for permits and concessions and of the relative delegated powers of the administrative authority.

4.2.2.3 POWERS OF THE PUBLIC AUTHORITIES

Law 613 of 1967 also controls the revocation of the permit or concession with reference also to acts concerning the protection of the sea. Article 41 states that the Minister for Industry, together with the Minister for the Mercantile Marine, having heard from the Technical Committee for Hydrocarbons, must declare that a permit holder has lost his right to the permit (following notification of the reasons and the fixing of a term for objections) if, among other things, he has not adhered to the requirements of the mining and marine authorities. The same procedure applies to concession holders in similar circumstances (article 43, no. 3) or if they do not observe other obligations imposed with the concession provisions which is also punishable by the loss of concession rights (article 42, no. 9).

Since installations for research and exploitation are subject to State law, the delegated powers attributed to State organs in matters of their competence are exercised by the organs responsible for that part of the coast nearest to the installation in question (article 49).

The immission of substances into the sea which, on the basis of article 15 of Law 963 of 1965 (law on maritime fishing: see 4.1.5), would constitute an offence, is permitted subject to special authorisation. The authorisation, however, does not exclude responsibility in the event of injury to private interests, punishable under the Penal Code (nor does it exclude civil responsibility), in line with the general principles governing authorised activities.

4.2.3 Pollution from navigation

4.2.3.1 DISCHARGE OR LEAKAGE OF HYDROCARBONS FROM SHIPS

This matter is regulated at the international level by the London Convention of 1954 and amendments adopted in 1962, 1969 and 1971. These regulations are due to be replaced by the London Convention of 8 November 1973 for the prevention of pollution from ships, which should come into force 12 months after it is signed by 15 countries representing at least 50% of the world's merchant fleet.

COASTAL WATERS AND OPEN SEAS

Italy has put the Convention of 1954 into effect with Law 238 of 1961, and the amendments of 1962 by Law 94 of 1970, those of 1969 by Law 341 of 1974 and those of 1971 with Law 875 of 1975; the Convention of 1973 has not yet been put into effect in Italy.

The Geneva Convention on the High Seas, article 24, which has been put into effect in Italy, also mentions the subject of pollution from shipping.

The regulations in Italy are as follows: the discharge of hydrocarbons or mixed hydrocarbons into the sea from ships or tankers in any area is prohibited (the amendments of 1969 have eliminated the idea of forbidden zones), unless certain conditions are satisfied: for ships which are not petrol tankers, the ship must be on course, the rate of discharge of hydrocarbons must not be more than 60 litres per mile, the hydrocarbon content in the mixture discharged must not be more than 100 ppm of the mixture, the discharge must be carried out as far as possible from the coast; for petrol tankers: the petrol tanker must be on course, the rate of hydrocarbon discharge must not be more than 60 litres per mile, the total quantity of hydrocarbon discharged in a voyage or as ballast must not be more than 1/15000 of the total capacity of the ship, the petrol tanker must be more than 50 miles from the nearest land. Article 6 of the Convention states that the sanctions which can be applied by the signatory States in exercising a discharge prohibition for offences outside their territorial waters must not be less severe than sanctions for offences committed within territorial waters.

To facilitate controls all petrol tankers and ships must be equipped with a hydrocarbons register in which all loading, transfer and unloading operations for hydrocarbons transported must be registered, including ballast and discharging of ballast waters, tank rinsing waters, discharges of residues etc. With regard to port equipment required under the Convention to handle oily sludges, ballast and rinsing waters from ships which would otherwise discharge these substances at sea, see 4.1.4. The mechanism provided for under the Convention, to control infringements of the provisions, is based on collaboration between the governments participating in the agreement. Each government is entitled to supply to the government of the country in which a ship is registered a written communication denouncing contraventions of the Convention for which the ship itself is responsible, regardless of where the contravention occurs. The government receiving such a communication can request further details concerning the contravention. When the government in whose territory the ship is registered has sufficient proof for proceedings under its own legislation to be initiated against the owner or captain of the ship, it will make sure that such proceedings are initiated as quickly as possible and will notify the denouncing government and IMCO of the results of the proceedings.

The application of this legislation from an international source makes it possible to take action against pollution acts committed by foreign subjects, which occur in the proximity of Italy but outside her territorial waters. In the event of pollution in territorial waters or even in the open sea, but caused by Italian subjects (either as the material perpetrators, or as principle: article 15, f), Law 963 of 1965 applies since it guarantees a more immediate protection.

If the London Convention of 1973 for the prevention of pollution from ships was put into effect, it would introduce considerable amendments to the legislation described above. In fact this Convention adopts and reinforces the requirements of the previous international legislation and extends it to all forms of pollution from shipping. The Convention introduces the concept of special areas and among these areas is the Mediterranean. Consequently, in this area the discharge into the sea of hydrocarbons must be totally forbidden. The Convention of 1973 also provides for installations in ports and on the coast which are suitable for the handling and treatment of rinsing and ballast waters which the ships which arrive at ports or loading points must carry on board (see 4.1.4). Violations of the Convention are punished by the State in whose territorial waters the violation occurs or by the State whose flag the ship flies and the signatory States are obliged to bring their internal legislative systems on the subject up to a similar standard.

4.2.3.2 NUCLEAR PROPELLED SHIPS AND SHIPS CARRYING RADIOACTIVE SUBSTANCES

This matter is the object of the two international conventions for the safeguarding of life at sea (SOLAS) of 1960 and 1974. The first is fully in force in Italy, but Italy is not a party to the second. In addition to the crew and the passengers, the Convention of 1960 deals with protection of the international waterways and marine resources against radiation; the planning and construction of nuclear reactor installations must be controlled and approved by the competent authorities (see Chapter 7); the packaging, labelling and stowage of dangerous materials are subject to rigorous controls.

4.2.3.3 SHIPS TRANSPORTING HARMFUL SUBSTANCES OTHER THAN HYDROCARBONS

Here we must consider not only the discharges controlled under Law 319 of 1976 and by the Conventions of London of 1972 and Barcelona of 1976 (see 4.2.1) which are subject to authorisation procedure, but also discharges, dumping and emptying operations from shipping activities. In this context the provisions of the Navigational Code are applicable, such as article 71 prohibiting the dumping of any kind of matter, when this prohibition has been extended by the Head of the Maritime Department to cover areas other than the port (the area to which it normally applies). Article 15 of Law 963 of 1965 on fishing in maritime waters is also applicable: under this article a pollution act involving substances which alter the normal biological balance of the sea water constitutes an offence.

However, these provisions are only applicable for acts carried out in territorial waters or, under Law 963 of 1965, to acts carried out in international waters by Italian subjects.

The London Convention of 1973 (not ratified but signed by Italy) is also concerned with this type of pollution so that when it is put into effect it will provide further protection possibilities. The Mediterranean is considered a special zone for the purposes of the Convention also with regard to the release of wastes by ships (for which there are rigorous prohibitions and controls in force). The Mediterranean is, however, not considered a special zone (unlike other enclosed seas such as the Black Sea and the Baltic Sea) by the Annex concerning the release into the sea of chemical products used by ships.

4.2.3.4 POLLUTION CAUSED BY SHIPPING ACCIDENTS

The problem of preventing shipping accidents is handled by Law 616 of 5 June 1962 on the safety of shipping and human life at sea, and at international level by the SOLAS Convention of 1960 approved by Law 538 of 26 May 1966 (amendments were approved by Law 1235 of 19 November 1968). The second SOLAS Convention of 1974 has not been signed by Italy. Furthermore, the international controls for the prevention of accidents also include international rules for the prevention of collisions at sea, adopted in 1960, which came into force in 1965 and were modified in 1972; these international controls have been accepted in Italy.

Under Law 616 of 1962 Italian ships or ships entering Italian ports must conform with certain characteristics and stipulations relating to the safety of human life at sea. This observation of the regulation has to be certified by various documents (safety certificate, safety certificate for military or naval armaments, radiotelegraphic or radiotelephonic safety certificate, fitness certificate etc.). The navigability certificate is also of importance for ships of a gross tonnage of at least 25 tonnes not equipped with a first class certificate; the certificate for the free board load line determining the maximum load, the free board load line and the water line which are compulsory for ships with a tonnage of 150 tonnes or more intended for the international transport of cargo or those for the internal transport of passengers, or of at least 500 tonnes for the internal transport of cargo. Every port has an inspection commission nominated and presided over by the harbourmaster for the purposes of verification required under the law (except those relating to the free board load certificate which are the subject of a special regulation: DPR 579 of 13 March 1967). The law also provides penal sanctions for the captain of the ship or the shipowner for violating the safety provisions (loading in excess of the limit; abusing the free board load line; operational omissions; failure to notify the authorities of alterations carried out to the ship or of damage; failure to carry out or incorrect carrying out of safety works relating to navigation; failure to comply with approved characteristics).

International instruments to reduce to the minimum damage to the marine environment caused by possible navigational accidents have been accepted and put into effect in Italian legislation. Thus, the amendments to the London Convention of 1954 on oil pollution, adopted in London in 1971 and put into effect in Italy by Law 875 of 19 December 1975, provide construction standards for petrol tankers, particularly with regard to the size and characteristics of their tanks. A more comprehensive and rigorous control extended to all types of maritime transport and concerning construction, inspection and certification, is contained in the London Convention of 1973 on the prevention of pollution from ships. This Convention has not yet come into effect but it is expected to replace the international regulations currently in force.

With Law 185 of 6 April 1977 and DPR 504 of 27 May 1978 respectively, Italy has ratified and rendered effective the International Convention on Intervention on the High Seas in cases of Oil Pollution Casualties, signed in Brussels in November 1969, which authorises countries threatened or affected by hydrocarbon pollution as a result of a shipping accident to take every necessary measure to avoid or reduce damage, including the destruction of the ship or its cargo. By these means even interests connected with the safeguarding of the sea and the coastline can be protected (i.e. the conservation of the biological resources of the sea, of aquatic wildlife, fish etc.).

For pollution from the same cause but by substances other than hydrocarbons there is a protocol on intervention at sea in the event of an accident which leads or could lead to pollution by substances other than hydrocarbons, adopted in London in November 1973, recently ratified (see 4.1).

4.2.4 Contingency plans in the event of accidents causing pollution from hydrocarbons

As is well known, intervention or contingency plans are aimed at ensuring a prompt and adequate response in the event of pollution arising from any accident at sea during the course of any activity. These plans should specify the authorities responsible, which forces should intervene, the intervention procedure, the technical means available (crew, materials, ships, aeroplanes, helicopters) and coordination.

The plan should also consider the measures to be taken to protect human beings, flora, fauna and, in general, natural resources, commercial, touristic and recreational interests.

In 1972 the Ministry of the Mercantile Marine formulated a national emergency plan to combat pollution caused by hydrocarbon spillages at sea following accidents. This operational plan does not interfere with the provisions and regulations in force concerning maritime disasters, nor with other legal provisions. It applies when a hydrocarbon pollution accident is of such magnitude that it cannot be eliminated by the anti-pollution means available in the areas affected.

The plan considers all types of accident at sea which could produce hydrocarbon pollution or a potential threat of pollution, such as:

maritime disasters involving one or more tankers (collision, explosion or other causes);

maritime accidents involving steamships which can result in danger of pollution caused by hydrocarbons;

pollution caused by hydrocarbons from accidents on stationary or mobile installations situated in territorial waters, on the coast or the continental shelf.

The local maritime authorities are appointed to intervene in any event as a preliminary action with all the means at their disposal while waiting for the national plan to be put into effect. A permanent coordinating committee has been set up to ensure the coordinated intervention of the various civil and military forces which have suitable means to combat the pollution.

The head of the maritime district of the area adjacent to the afflicted zone is responsible for implementing the plan.

The coordinating committee is presided over by the Director General of the Seaboard, Territorial Waters and Ports of the Ministry of the

Mercantile Marine. Representatives from the following offices and ministries are members of the committee: Marine General Staff; Aeronautical General Staff; Ministry of the Interior; Public Security General Directorate; General Directorate of Civil Protection and Fire Services; General Command of the Carabinieri; General Command of Customs Officers; Italian Naval Register; National Hydrocarbon Organisation—ENI; Petroleum Union. The head of the maritime district responsible for operations can make use of the cooperation of a group of technical experts.

The coasts and sea water under national sovereignty have been assessed and classified, on the basis of three criteria, for the purpose of creating a comprehensive and well distributed system of protection in the event of a national emergency: with regard to the intensity of petrol tanker traffic; with regard to the proximity to the coast of the navigation lanes used by petrol tankers; with regard to the intrinsic value of the various zones as national resources, touristic and recreational areas.

The intervention plan contains a continuous monitoring system which is of particular importance bearing in mind the characteristics of the Mediterranean and the morphology of the Italian coast which make rapid intervention necessary in order to prevent the hydrocarbons reaching the coast. Other technical intervention systems are also provided for, depending on the magnitude of the pollution accident and the prevailing conditions at sea.

Finally, coastal defence measures can be put into effect in the event of intervention at sea being unable to prevent the danger of coastal pollution as a result of atmospheric conditions.

Intervention centres strategically placed in carefully selected ports hold ships, equipment and anti-pollution material available for use within an 80 mile range.

Decree 886 of the President of the Republic of 24 May 1979 on exploration, research and exploitation of liquid and gaseous hydrocarbons in territorial waters, the continental shelf and other submarine areas which come under the powers of the State, is also of relevance with regard to intervention and safety at sea from pollution caused by hydrocarbons.

This decree incorporates provisions on the same subject contained in Decree 128 of the President of the Republic dated 9 April 1959, with special provisions to prevent pollution, protect the safety and health of workers, avoid damage and danger to other legitimate uses of the sea, to the marine fauna and flora, and to submarine pipelines, cables and installations. For each installation there must be an emergency plan which includes instructions for the conduct of all personnel, to be rehearsed periodically, as specified in the emergency plan. If, in spite of the preventive safety measures and precautions, spillages of oily substances into the sea occur, immediate steps must be taken using the most suitable methods available to contain, remove or neutralise the polluting substances. Equipment for any necessary intervention must be held available on the platform, support ships and on land, in accordance with provisions to be issued by the Ministry of the Mercantile Marine and the Ministry of Industry, Commerce and Craft, within 3 months of the entry into force of this law (provisions not yet issued).

In the event of a serious accident the person in charge of the platform should request the intervention of the Maritime Authorities. At international level, a protocol has been signed and ratified by Italy under the Barcelona Convention of 1976, concerning regional cooperation between the various Mediterranean States, to combat pollution caused by hydrocarbons and other harmful substances.

A regional coordination centre has been set up in Malta by the Mediterranean States. The protocol requires the contracting parties to take the necessary measures, in the event of serious and imminent danger to the marine environment, to maintain and promote intervention in the event of emergency (contingency plans) individually or through bilateral or unilateral cooperation; to disseminate information; to exchange information in the event of an emergency, either directly or through the regional centre; to assist each other in combatting pollution from hydrocarbons or other harmful substances; and if necessary, to coordinate operations through the regional centre.

In addition, monitoring activities with regard to acts of pollution are envisaged to be carried out through bilateral or multilateral collaboration with the help of ships and aeroplanes registered in their territories, and accidents, hydrocarbon leakages, sightings of oil slicks or other polluting phenomena are to be notified.

Collaboration on the subject of contingency plans already exists through bilateral conventions between Italy and Yugoslavia, Italy and Greece, and Italy and France.

4.2.5 Civil liability for pollution of the open sea

4.2.5.1 POLLUTION CAUSED BY SHIPPING

There are no specific regulations on this subject in Italian legislation: the general regulations contained in the Civil Code for unlawful damage can be applied. We should remember that the system of the Italian Civil

Code can provide protection for individual situations and interests of an indefinite number and nature, provided that the injury can be considered 'unlawful'. Italy has not ratified the International Convention on the Limitation of Liability of owners of sea-going ships, signed in October 1957, which does not especially refer to pollution damage, but nevertheless applies to it.

Law 185 of 6 April 1977 and DPR 504 of 27 May 1978 have ratified and implemented the Brussels Conventions:

(i) International Convention Relating to Intervention on the High Seas in Cases of Oil Pollution Casualties, with annex, adopted in Brussels on 29 November 1969.

(ii) International Convention on Civil Liability for Oil Pollution Damage, with annex, adopted in Brussels on 29 November 1969.

(iii) International Convention on the Establishment of an International Fund for Compensation for Oil Pollution Damage, adopted in Brussels on 18 December 1971.

There is nevertheless continuing affirmation of the idea of a system of civil liability ever more centred on objective criteria of imputation of responsibility, especially in the field of economic activities: on the one hand the idea of responsibility as a 'cost' is favoured, and on the other hand the need is felt for a rationalisation and control of the damage produced by activities with a high level of social danger, utilising the regulations for civil liability with ever more emphasis on compensation rather than 'punishment' for acts resulting in unjust damage. Above all, the importance emerges of regulations such as those on damage from dangerous activities (article 2050, Civil Code) which places an objective responsibility on the person carrying out a dangerous activity which can result in damage if the proper precautions are not observed (i.e. all proper, technically possible measures, even if they are very costly and uneconomic).

However, the custom is widespread of stipulating insurance for damage caused to third parties by objects or substances discharged by ships, whereas for aircraft insurance against damage from dumping or loss of materials is obligatory.

4.2.5.2 POLLUTION CAUSED BY OFFSHORE INSTALLATIONS

The considerations of 4.2.5.1 can be repeated here, both on the lack of specific regulations, and on the methods and characteristics of recourse of the Civil Code system.

At international level, in the northern European States the problem has been dealt with by the Intergovernmental Conference on the Convention on Civil Liability for Oil Pollution Damage Resulting From Exploration for and Exploitation of Sea Bed Mineral Resources (London, December 1976). According to the text of the Convention, the permit-holder can release himself from responsibility only by demonstrating that the damage was caused by an act of war, civil war, insurrection or by an inevitable and irresistible natural phenomenon of exceptional character. Limitations of responsibility and requirements are imposed regarding the joint responsibility of more than one operator involved in the same installation.

In the Mediterranean there is as yet no convention on the subject. The problem, deeply felt and the object of research, is one of the subjects for future discussion among the Mediterranean States of the Barcelona Convention, for the purpose of arriving at comprehensive and harmonised legislation on the subject at both the international and national levels.

5
Deposit of Waste on Land

In this chapter we shall deal both with legislation concerning the collection and disposal of municipal solid wastes and with the legislation contained in Law 319 of 1976 (see Chapters 3 and 4) concerning the discharge of sludge and effluent on land.

The legislation concerning municipal solid wastes is contained in Law 366 of 20 March 1941 and subsequent amendments. This law distinguishes between 'external' wastes (arising from discharges carried out in public areas) and 'internal' wastes (produced by offices and homes) and also includes provisions concerning the recycling of usable wastes in industrial manufacturing industries or agriculture (even though these provisions have remained more or less ineffective).

The collection and disposal of municipal solid wastes is undertaken by the communes, either directly or by means of a concession granted to a private contractor. The communes meet the costs of this service with the proceeds of an *ad hoc* tax proportional to the premises occupied by the taxpayer and to the use to which the premises are put.

Some rules regarding the way in which the service is carried out are laid down by the law: wastes must be removed in hermetically sealed transport containers; wastes must be collected daily and not left at the collection point; the time during which collection vehicles are stationary in urban centres must be kept to a minimum etc.

The communal regulations complement the requirements of the law indicating precise controls for the service.

Supervision and control are the responsibility of the Ministry of Health, which issues general instructions on the way in which the service should be conducted.

In fact, the law made provisions for the constitution of other organs with control functions over the collection and disposal service, and also for the promotion of studies and research (such as the Central Office of the Ministry of the Interior, which was responsible for this sector until the constitution of the Ministry of Health in 1958); provisions for organs with consultative functions were also made (like the Central Commission), none of which, however, was ever set up (likewise, the register of companies qualified to manage municipal cleansing services was never instituted).

The lack of application, by now hardly surprising, of the regulations for the reutilisation of wastes and the difficult financial situation of the communes which weigh heavily on the efficiency of the service are other weak points in the control system. The question of the disposal on land of effluent from discharges arising from manufacturing and civil installations, and residual sludge from industrial processes and purification processes, is controlled by Law 319 of 1976 and is the object of detailed and technical requirements dictated by the Committee of Ministers (now the Interministerial Committee) under the law (article 2, *e*), in the resolution of 'general technical provisions' of February 1977 (Annex 5). The law distinguishes between the discharge of effluent on land (also permitted for agricultural use provided that the effluent is of direct benefit to agricultural production), disposal of residual sludge from industrial or purification processes, disposal, through plants, of effluent discharged on land and arising from civil installations of less than 50 rooms or 5000 cubic metres. In addition Table B annexed to Law 319 of 1976 appeared to apply to discharges on to land or into the subsoil, but the Table was never specifically referred to in the text of the law itself with the exception of article 9; in any case, the table was abolished by article 23 of Law 650 of 1979.

The control of discharges on to the land or into the subsoil is the responsibility of the communes, consortia and mountain communities. The public disposal services for residual sludge from manufacturing processes or effluent treatment plants are managed, in addition to the above authorities, by consortia instituted by the special statute regions or by consortia for the areas and industrial development centres provided for under the TU of the laws relating to the Mezzogiorno (DPR 218 of 6 March 1978).

In addition to their management tasks, the regions are also responsible for coordination and planning of the whole question of discharges, for issuing integrative and implementing rules for the general technical provisions laid down by the Interministerial Committee for the disposal of sludge and effluent, and in particular the delimitation of areas in which the disposal of effluent on land or into the subsoil is permitted (article 4 (*e*) of Law 319 of 1976).

5.1 CONTROL OF DEPOSITS OR DISCHARGES

5.1.1 Locations for deposit or discharge

Under Law 366 of 1941 it is forbidden to discharge, accumulate or deposit municipal solid wastes on public or private lands in a city. There is also an obligation to avoid the accumulation of waste inside courtyards, entrance halls or passages of public or private buildings.

Areas for the disposal of wastes are carefully selected by the mayor in consultation with the Health Official, with the approval of the Provincial Medical Officer who is aided by the Provincial Commission for Solid Wastes. The disposal plant must be sited at least 1 kilometre from the nearest dwelling.

'External' wastes, i.e. those produced by industrial or similar activities and discharged in public areas, can be removed by the commune (outside the normal municipal cleansing service or private contract for this service) which provides for this by a decision of the communal council, making arrangements for the methods and charges for removing such wastes.

Article 217 of the TU health laws also applies to industrial wastes, authorising the mayor to issue regulations and, in the case of non-compliance with these regulations, to intervene in the event of danger to public health resulting amongst other things from solid wastes arising from manufacturing processes or factories.

With regard to the discharge of effluent and sludge on to land we have already mentioned the regional responsibility for delineating the areas where the disposal of such wastes is permitted within the framework of the regional powers to integrate and put into force the general technical provisions. Discharges on to land used for agricultural purposes are only permitted if the wastes are of direct use in agricultural production (the practice of locating discharge and sewage pipe outlets on horticultural or agricultural fields frequently encountered in the past and substantially correct is now in crisis, due to the presence of detergents, synthetic foam, non-degradable substances and wastes from industrial production in the effluent, which has resulted in the impoverishment and degradation of the land). Discharges into the subsoil should not be permitted if they can damage the groundwaters.

Annex 5 to the resolution of the Committee of Ministers of 4 February

1977 defines the characteristics of the site where the disposal of effluent or sludge can take place. Important factors to be considered are temperature, rainfall, humidity, wind, vegetation and relative transpiration—evaporation. The nature of the geological structure should be investigated and also the depth, profile, structure, texture and conduction abilities of water and electricity of the land etc. For sites used for agricultural purposes, the gradient of the slope and the nature of the land should be such as to avoid flooding problems (also with the aid of adequate water systems). There are particular requirements to avoid the contamination of existing groundwaters. The area must also be surrounded by a strip of land 80 metres wide on which no buildings, national or provincial roads must be constructed. Suitable placards must be erected to designate the area and indicate any eventual hygiene risk.

The discharge of effluent and sludge arising from industrial processes or purification plants onto land is forbidden subject to hydrogeological restrictions. Disposal sites for sludge must be approved by the responsible authority. If alternative sites are not available and there is a lack of public services for the disposal of sludges, the authority must identify areas where the sludge can be disposed of in accordance with the law and technical regulations. The site for sludge disposal must also be selected according to criteria relating to climate, land characteristics, topography, geological and hydrogeological conditions.

5.1.2 Method of deposit or discharge

Law 366 of 1941 requires solid urban wastes to be disposed of in suitably controlled areas, or alternatively they can be incinerated. Disposal on land is carried out by landfill (infilling of natural or artificial holes or cavities) or by controlled tipping (layers of waste interspersed with layers of earth). Other methods of waste treatment in use in some communes are the transformation (total or more often partial) of the wastes into fertilizer or compost. Tipping in specified areas (but unfortunately often not controlled areas) remains the most popular method, followed by incineration or burning of waste in the open (there are as yet few incinerators or transformation plants for turning the wastes into fertilizer). One of the most difficult problems in this respect is the communal budget: the waste collection and disposal service constitutes a considerable item of expenditure in the communal finances and collection and transport uses up most of this sum (three-quarters on average), leaving little money available for the provision of efficient disposal plants. There is no precise technical regulation on the disposal of solid municipal wastes, nor on the methods of discharge and disposal most suited to the various types of waste; everything is left to the provisions of the local authority, communal regulations, orders from the mayor and health regulations from the Provincial Medical Officer.

There are, however, detailed technical provisions concerning the discharge and disposal of effluent on land, or of sludge arising from industrial or purification processes, or effluent from civil installations of less than 50 rooms or 5000 cubic metres.

Effluent discharged on land used for agricultural purposes and cultivated with products intended for human consumption in the raw state must be subjected to primary and secondary treatment (and, if necessary, filtration and disinfection). In the case of products intended for human consumption after cooking or other processing, or intended for animal foodstuffs, the treatment, filtration and disinfection are required to achieve only a lower level of bacterial purification.

Discharge on land not used for agricultural purposes must respect the interests of the landscape, nature and land conservation. For all discharges, the distribution of effluent on land must be controlled by means of sprinkling, lateral infiltration, submersion, spraying and so on, depending on the condition of the land, the extent of the discharge, the nature of the vegetation, etc. However, the way in which the discharge is carried out must be compatible with the need for stability and efficiency of the natural purification processes required. It is also necessary, in the areas where disposal on land is permitted, that the transport and distribution of effluent should take place through enclosed channels and that on land used for agricultural purposes there should be collective distribution plants. Other requirements concern techniques for facilitating the disposal and break-down of the organic substances as well as inspections and controls to ensure the efficiency of the plant.

Sludge arising from industrial or purification processes must undergo appropriate treatment to make it suitable for its chosen destination, while at the same time safeguarding the possibility of recovering any reusable substances it may contain (there are also provisions for the constitution of consortia for the treatment of sludge and the recovery and recycling of substances or of economic or energy value contained therein). Sludges with a prevalently organic content must undergo at least one stabilising treatment to reduce the presence of pathogens and noxious odours. It is also required that, in cases where the sludge cannot be rendered totally harmless, it be subject to controlled storage. The application of sludge on to land can be carried out in its liquid state or after dehydration. It can be dispersed by spraying, spreading or depositing on to the surface layer, depending on the characteristics and use of the site (the sludge must never come into direct contact with foliage in plantations). The application of liquid sludge must be suspended if it is not sufficiently absorbed by the terrain. Dehydrated sludge is applied either by simple accumulation, or spread over the land, or by layering with alternate layers of earth, or by filling in trenches to be subsequently covered with earth. In the event of surface spreading or accumulation, the dispersion of substances into the atmosphere must be avoided by using windbreaks, grass coverings, etc.

There are certain obligatory types of plant for the disposal of effluent to be discharged on land from civil installations of less than 50 rooms or 5000 cubic metres. The authorised methods fall into two categories:

(i) accumulation and fermentation in sewage receptacles with periodic extraction of the material and its burial in a manure heap, or other suitable disposal;

(ii) clarification (in a traditional septic tank or one of the Imhof type) followed by oxidation for dispersion on the land through sub-irrigation or absorbent wells, or soak-aways into the ground through sub-irrigation by means of drainage. For each of these methods there are specific conditions and characteristics.

All the requirements referred to above with reference to the disposal of effluent and sludge are contained in the resolution of the Committee of Ministers (under Law 319 of 1976) adopted on 4 February 1977. The regions have a responsibility for the integration and implementation of these general technical regulations.

5.1.3 Controls over the quality of the discharge

Provisions concerning the qualitative characteristics of discharges which contaminate the land exist only with regard to the discharges of effluent or sludge controlled by Law 319 of 1976. Discharges from new manufacturing installations are by law required to observe the provisions issued on the subject by the State (Interministerial Committee), the regions (in their function of integration and implementation), and the hygiene provisions laid down transitionally by the local health authorities. Manufacturing installations already in existence when Law 319 of 1976 was issued must make their discharges on land conform to the provisions established by the local health authorities while at the same time ensuring that they are in line with the permitted limits under Tables A and C annexed to the law (see Chapter 3), and subsequently that they observe the specific provisions laid down by the Committee of Ministers and the regions (responsible for the definitive approval for the discharge). Under article 14 of Law 319 of 1976 discharges on to land from civil installations must be controlled by the regional treatment plans. Nevertheless, while that is valid for existing discharges, it would be more in line with the system of the law if new discharges, right from the start, were subject to permitted limits. Moreover, before the definition of the regional plan, the regions must adopt the necessary measures to protect public health through their health organs (article 14, new text, Law 319 of 1976).

The general technical provisions decided on 4 February 1977 by the Committee of Ministers contain precise indications concerning the qualitative parameters of the discharges. Discharges of effluent on land must comply with Table A of the law, and in addition there are further parameters: permeability, concentrations of sodium, calcium, magnesium, etc. The discharge should not contain substances which alter the structure of the soil and in particular the permeability and aeration. The quantity of organic substances applied must not exceed the cleansing capacity of the soil. Liquid discharges must not contain organic substances which are not easily biodegradable. It is stated that the levels given in Table A for some substances are not sufficiently cautious for discharges on to agricultural land.

Sludge deriving from manufacturing or purification processes must not contain substances which alter the chemical and physical structure of the soil. The maximum permitted concentration of persistent or bioaccumulable toxic substances must be assessed in relation to the characteristics of the land and the state of the substances. Furthermore, the technical provisions set out a series of examples of parameters for assessment of the characteristics of sludge in relation to the substances listed in the following table:

pH	Potassium	Total coliforms and pathogens
BOD_5	As, Cu, Cr, Cd, Hg, Ni, Pb, Se, Zn, B	Specific weight of solids and solids granulometry
COD	Oils and fats	
Organic carbon	Hydrocarbons	Conductivity of aqueous extract
Ammoniacal, nitrite, nitrate and total nitrogen	Surfactants	SAR index of aqueous extract
Chlorides	Toxicological characteristics	Dry residue at 105 °C and 600 °C
Fluorides	Biocides	
Total and soluble phosphates	Chlorinated organic substances	
Soluble silicates		

Promulgation of a law is awaited as delegated to the government by Law 42 of 9 February 1982, with regard to the implementation of Directive EEC/439/75 regarding the disposal of used oils. (Chamber Act 1903, see 7.2.3.1.)

5.2 POWERS OF THE PUBLIC AUTHORITIES

5.2.1 Administrative powers

The regions have been responsible for protection of the land from pollution (and thus control of the collection and disposal of solid municipal wastes) since 1974 (DPR 4 of 1974). The Provincial Medical Officer is in charge of regulatory management and control of the communes, with reference to running the urban cleansing service. This protection of the land is one of the most effective sectors of intervention within the scope of regional legislation on environmental matters.

Discharges on to land of effluent or sludge are subject to authorisation under Law 319 of 1976 and must comply with the requirements and limits fixed by that law (see 5.1.3). The authorisation is granted by the commune, consortium or mountain community and its procedure is regulated by the possibility of provisional regulation and by the possibility of repeal under the general regulations of Law 319 of 1976 (see Chapter 3).

5.2.2 Judicial powers

Violation of the laws on municipal solid wastes is punishable as a contravention under the Penal Code. The violation of prohibitions imposed by the law, principally concerning the discharge of waste, is punishable by a fine of up to 80,000 lire. Article 674 of the Penal Code applies, in addition, to the above-mentioned offences. Under this article the depositing or discharging of waste on land which results in damage or danger constitutes an offence.

The sanctions provided for under article 21 ff. of Law 319 of 1976 apply to discharges of effluent or sludge on land; failure to obtain an authorisation or non-observance of the requirements indicated in the authorisation provision or in the law can even result in detention under these sanctions (imprisonment of up to 2 years in the most serious cases—see 3.6).

5.3 RIGHTS OF THE INDIVIDUAL

See 1.4.

5.4 NEW LEGISLATION PROPOSED FOR THE DISPOSAL OF SOLID WASTES

Under the stimulus of Community legislation (EEC Directive 422 of 15 July 1975 and EEC Directive of 20 March 1978 78/319), bills of law have been prepared for the purpose of providing new regulations on the subject. In particular, a draft law has been presented by the Government (Senate document 1044, VII legislature, which expired at the end of this) and another has been formulated by the Senate Special Commission for Ecological Problems (Senate document 1441, VII legislature, which expired and was subsequently re-presented as Act 954, VIII legislature). Both these draft laws declare the activity of the disposal of wastes in all its various phases to be an activity of public interest, and they dictate general principles regarding methods and aims (above all the protection of health, the countryside and the environment). A classification of wastes into two categories is introduced: municipal wastes and special wastes. The former are those arising from buildings or civil installations, used and abandoned consumer durables, and street litter. The second category includes wastes from industrial and agricultural production, or from craft and commercial activities similar to these; wastes arising from hospitals, nursing homes, etc.; demolition, construction and excavation materials and dilapidated or obsolete machinery and apparatus; abandoned cars, trailers, etc.; sludge arising from effluent treatment and residues from waste disposal activities. The activities of municipal waste disposal and the disposal of special wastes consisting of sludge from purification processes are the responsibility of the communes, whereas the disposal of other special wastes (including the cost) is the responsibility of the manufacturers of the waste itself, who can, however, make use of the private contractors authorised by the regions, or else entrust their waste to the commune on the basis of a suitable arrangement between the parties concerned. The commune can in turn manage the service itself, through a municipal agency, or alternatively by contracting a specialist company authorised by the region to carry out the service on its behalf.

Appropriate communal regulations determine the hygiene standard to be observed, in addition to the requirements for the recovery of reusable materials, the area within which the service is to be carried out, the methods to be adopted, and the tax payable for the disposal of the wastes. The administrative system under the draft laws allows for wide regional responsibility (and the project of the Senate Special Commission also allows for provincial responsibility for the control of the application of the regional provisions on discharges and disposal installations). Above all, the regional powers of planning integrate and define the system of the law which for this reason assumes the properties of a framework law. The regions should prepare programmes for the organisation of services, approve projects for collection, discharge and disposal plants, formulate plans for financial intervention in favour of the communes, authorise specialist waste disposal companies, prepare periodic data and statistics on the subject, issue provisions for improving existing plants, etc. In addition, within the framework of regional programming, the constitution of mixed consortia is envisaged (involving communes and individuals producing special wastes).

The draft Government law also envisages the setting up of an Interministerial Technical Committee within the Ministry of the Interior (whereas the Senate draft law makes provision for the constitution of a proper Interministerial Committee). The Technical Committee should indicate the criteria and general technical provisions for the operation of the service, for the running of installations, for the permitted limits for toxic substances contained in the wastes, and any special disposal procedures. The Minister of the Interior implements these provisions by ministerial decree (the institution of an Office for Solid Wastes within the General Civil Administration Direction of the Ministry of the Interior is envisaged). The Government project also provides ordinance powers (in cases of necessity and emergency) for the prefect and mayor (exercised by arranging for particular forms of disposal).

In addition, EEC Directive EEC/442/75 relating to refuse and Directive EEC/319/78 relating to toxic and harmful waste await the promulgation of a law as delegated to the Government by Law 42 of 9 February 1982.

5.4.1 Municipal wastes

The draft Government and Senate laws impose general principles to be observed in the management service and to be integrated into the communal regulations. The prohibition concerning the dumping of wastes in public or private areas, in town roads, squares and public markets is reiterated. Particular attention is also given to the question of the containers to be used and to cumbersome wastes.

5.4.2 Special wastes

The disposal of wastes arising from industrial or agricultural or similar processes and demolition or similar wastes is the responsibility of the producer of the wastes and can be carried out by them or through authorised private contractors; alternatively, the producers can make a suitable arrangement with the commune for the disposal of the wastes (which, of course, is at the expense of the waste producers). Wastes arising from hospitals, nursing homes etc. should be incinerated on the spot in incineration plants which must conform to the requirements laid down by the technical interministerial committee, or else be transported in suitable containers to be incinerated in communal plants or installations managed by authorised contractors. Car bodies, trailers and similar objects should be collected in appropriate centres for breaking and recovery of usable parts and scrap metal (in areas selected by the region in consultation with the communes involved). The management of these collection centres should be authorised by the public safety authorities and licensed by the commune if the manager is not the commune itself or an intercommunal consortium. Incentives are envisaged for suitable industrial establishments to encourage the recovery of materials. The commune must be notified of the quantity and quality of sludge arising from the purification of industrial effluent and the system employed for its disposal.

5.4.3 Systems of disposal

The draft laws make a point of encouraging disposal systems which allow for recovery of the energy content of the wastes and which at the same time are convenient from a technical, economic and environmental point of view (taking into account the question of noise pollution which can arise from these plants: article 25 (*d*) of the Government draft law). The waste treatment method is chosen by the commune on a comparative basis with other possible methods and the plans for the installations are approved by the regions (these must be authorised by the commune if they are being installed by the producers of the waste they are to treat).

With regard to disposal methods, uncontrolled tipping is prohibited, whereas controlled tipping, processing, incineration, pyrolysis, compaction, waste recovery and other methods which can be authorised by the Interministerial Technical Committee, even in the experimental stage, are taken into consideration. For each of these processes there are characteristics and methods of adjustment, as well as definite time-periods with which existing plants must comply.

The promulgation of a law, as delegated to the Government by Law 42 of 9 February 1982, is awaited to implement Directives EEC/403/76 relating to the disposal of polychlorodiphenyls and polychlorotriphenyls, EEC/769/76 concerning the harmonisation of legislative, regulatory and administrative requirements of member states relating to restrictions on the marketing and use of certain dangerous substances and preparations, and directive EEC/176/78 relating to wastes arising from the titanium dioxide industry.

5.4.4 Powers of the public authorities

One of the problems which must be solved by the new legislation is the question of financing the waste disposal service. The existing draft laws are concerned on the one hand with strengthening the powers of the communes by more efficiently defining the waste disposal tax collected by the communes themselves, while on the other hand providing financing mechanisms for disposal plants through ministerial and regional loans and grants.

This, however, is the question most susceptible to change and review during the parliamentary stage of the draft laws, and it is therefore impossible to discuss them in any detail at the time of writing.

The system of sanctions for the non-observance of obligations under the law has also been reinforced. The Government draft law provides for provincial control over disposal installations (via the Provincial Laboratories of Hygiene and Preventive Medicine and an appropriate provincial commission). In addition, the Area Councils should control the way in which the service is run, on Area Councils (Law 278 of 8 April 1976). Fines are enforced by the Administration for violations of the legal requirements.

6
Noise and Vibration

There is very little specific information available regarding noise pollution. No specific law exists and none is in preparation. No significant provision has been issued by the regions. Noise pollution is not explicitly mentioned in delegated decrees, either among matters under State control, nor among those transferred to the regions.

The term 'acoustic pollution' can be found in only two recent regional laws: Law 49 Reg. Lombardia of 23 August 1974 and Law 52 Reg. Piemonte of 21 August 1978.

In article 6 of the above mentioned Lombardy regional law the regional council is authorised to buy 'equipment for the sampling and direct control of noise emissions'; and under article 9, clause 3 a sum of 700,000,000 lire is made available for buying such equipment. The Piedmont regional law is more complete and makes provision for the implementation of the law (see article 10) of a total expenditure in 1978–1980 of 1,800,000,000 lire, of which 600,000,000 lire was already appropriated for 1978. The Regional Committee Against Atmospheric Pollution, instituted under Law 615 of 1966 (in carrying out the measures under article 101 of DPR 616 of 24 July 1977), appears extended to include, as required under article 8 for the purpose of examining the problem of 'acoustic pollution':

- an expert in industrial chemistry;
- an expert in toxicology;
- two acoustics experts.

The technicians chosen by the regional council from experts in public research institutes are appointed for a term of 5 years.

Article 9 authorises the regional council, for the purposes of preventing this type of pollution, to arrange appropriate contracts with public research institutes specialising in the subject, in order to make available the elements necessary for future regional legislation on the subject (in this context there is a joint project in progress between the region and the Galileo Ferraris Institute).

For the rest, legal regulations can be used which, while they are intended for other purposes, can be of use in protecting the environment from noise pollution.

The Civil Code regulations on immissions (see 1.2.3.1 regarding their application to environmental protection) can also refer to noise, vibration, jolting or similar phenomena.

In the Penal Code there are provisions for protection from annoying noise; for example, article 659 (disturbance to the occupation or peace of individuals).

Provisions on public safety and health (for example article 66 of the TU public safety laws) can also be used and in general the regulatory powers of the local health authorities, or of the mayor with the assistance of the Health Official, can be considered if required for reasons of public hygiene.

The laws on the salubrity of the work environment take into account the noise factor so that in some particular environments the worker enjoys legal protection: DPR 547 of 1955 on the prevention of casualties at work (particularly articles 7 and 390), and DPR 303 of 1956 on hygiene in the workplace (above all articles 24 and 33, the latter with attached table) should be considered here.

With regard to mobile sources of noise pollution, there are various provisions under the TU 393 of 1959 of the road traffic laws (which provide controls at the approval stage of vehicles and requirements for their use) and in the subsequent implementing regulation (DPR 420 of 1959).

Knowledge about the problem begins to be sufficiently comprehensive for the purposes of adequate legislative intervention: for example, there is the study carried out by the Special Commission for Ecological Problems of the Senate of the Republic, 'Report concerning pollution from solid wastes, the safeguarding of humid areas and noise pollution' (communicated to the Presidency of the Senate on 13 December 1971, doc. xxv no. 1).

6.1 STATIONARY SOURCES

6.1.1 Control over the siting of noisy activities

There are no specific regulations on the subject but general instruments can apply: town planning, for example, can take the noise factor into account, particularly with regard to industrial installations and the plans for zones produced by industrial consortia. However, the regulation concerning noxious industries under article 216, TU health laws (see 2.2.1) applies to noisy industries. The criteria for applying this legislation have been specified in circular 162 from the Ministry of Health dated 23 September 1971 (which will be discussed under 6.1.3). The Council of State (15 April 1955, in *Giust. civ.* 1955, II, 145) had considered the application to noisy industries of the law relating to noxious industries by the communes.

There are some provisions for the acoustic insulation of buildings but these are restricted to subsidised and public buildings (hospitals, schools etc.). See circular 1979 of 30 April 1966 from the Ministry of Public Works.

6.1.2 Protection of workers against machine or plant noise

The only provisions imposing design or construction criteria on plant or machinery to reduce the noise danger are contained in the legislation on safety and health at work. Article 24 of DPR 303 of 13 March 1956 on health at work states that 'in processes producing jolting, vibration or noise which is harmful to workers, measures must be taken, on the advice of technicians, to reduce the intensity'. It should be remembered that article 7 of DPR 547 of 27 April 1955 (on the prevention of accidents at work) forbids the construction, sale or hire of machinery or the installation of plants which do not conform to the provisions for preventing accidents at work.

With regard to the health of workers exposed to noise or vibration, preventive measures relating to possible occupational disease are taken in the form of compulsory annual medical examinations, as shown in the table attached to article 33 of DPR 303 of 1956.

It should be noted that article 9 of the Workers' Statute (Law 300 of

1970) gives workers' representatives not only the right to control the application of safety regulations for the work environment but also the power to promote the research, formulation and implementation of suitable measures to protect the health and physical wellbeing of workers.

6.1.3 Emission limits

The Ministry of Health circular 162 of 23 September 1971 has provided indications of a general character on the basis of article 216, TU of the health laws.

The following maximum limits have been set by the Ministry with regard to noise caused by industrial activities:

60 decibels B for day-time emissions;

40 decibels A for night-time emissions.

A tolerance of 20 decibels is permitted, in addition to background noise, with a maximum of 40 decibels A during the night and 60 decibels B during the day when the background noise is below these levels. However, the application of these limits is made somewhat difficult because of the lack of knowledge concerning noise levels in industrial areas.

There are also limits of an optional nature which are utilised by decisions of the courts as criteria to assess the illegality or otherwise of noise immissions under article 844 of the Civil Code (for illegality under this article the immissions must be 'intolerable'). These criteria are very elastic, depending on the actual situation (the regulation itself takes into account the conditions of the location); for residential buildings the tolerability limit is considered to be 40 phon for bedrooms and 45 for living rooms (Trib. Torino, 9 January 1957, Trib. Cuneo, 29 July 1958).

6.1.4 Powers of the public authority

In addition to powers concerning the control of noxious industries (see 2.2.1) there are two further types of power: the power of the local authority to issue regulations or ordinances, and the power concerning the penal repression of certain acts which constitute acts of acoustic pollution.

6.1.4.1 ADMINISTRATIVE POWERS OF THE LOCAL AUTHORITY

The municipal regulations issued by the communes usually control noisy activities, making them subject to restrictions issued on a case-by-case basis by the authority; insulation of rooms in which noisy activities take place can also be required. Violation of the regulations on noisy activities constitutes a contravention (article 659, clause 2, Penal Code).

A further important regulation is provided by article 66 of the TU public security laws, which requires the exercise of noisy professions or occupations to be suspended during certain hours, determined by local regulations or ordinances issued by the mayor. These regulations and ordinances (and therefore the offences committed if these are violated) concern continuous activities and not single episodes of acoustic pollution.

6.1.4.2 APPLICABLE PENAL PROVISIONS

We have already mentioned article 659 of the Penal Code. Under clause 1, an act by 'whoever, through noise or uproar, by abusing sound instruments or audible warning devices, or by inciting animals or failing to prevent them from making noise, disturbs the occupation or rest of others' is punishable by imprisonment of up to 3 months or a fine of up to 120,000 lire. As can be seen, this applies to noises which do not arise from continuous activities (unlike clause 2 which covers noisy activities contravening the requirements of the law or the authorities), but occasional acts. Under clause 1 and clause 2 two types of assessment are required from the judge; only in the case of clause 2 are manufacturing or business activities taken into consideration when the judge must assess the objective violation of the laws or regulations. The following judgements have been given under clause 2: Cass. 7 November 1975, Cass. 6 October 1975, 796; Cass. 2 July 1975, 797; Cass. 6 June 1975, 798 (all in *Mass. Cass. pen.* 1975, 795).

6.1.5 Rights of the individual

See 1.4 in general.

See 1.2.3.1 regarding the rights of individuals to take action against people producing immissions on the basis of article 844 of the Civil Code.

6.2 MOBILE SOURCES

6.2.1 Motor vehicles

The provisions regarding motor vehicles under the road traffic laws TU no. 393, 15 June 1959 and its implementing regulation DPR 420 of 30 June 1959 are more detailed.

6.2.1.1 CONTROLS OVER DESIGN, MAINTENANCE AND USE

All motor vehicles, motorcycles and mopeds must be fitted with an audible warning device; motor vehicles used for public transport must be fitted with a special audible warning device; motor vehicles used by the police or fire service must have an additional siren (article 46, TU).

Motor vehicles and mopeds must also be fitted with a suitable silencer to reduce the noise from the engine (article 47, clause 1, TU).

There is a dual control when the vehicle prototype is approved under article 53 on the noise emission level of the engine and the suitability of the warning device. In addition, the Civil Motoring Inspectorate performs an important role in carrying out the roadworthiness inspection (article 54) and subsequent overhauling of the vehicle (article 55).

The same TU under article 112 requires all vehicles to be fitted with a silencer which must be kept in good condition and not altered. Article 113 prohibits the use of audible warning devices in built up areas except in cases of immediate danger, or emergency transportation, and in any event such warning devices should be used in moderation even outside built up areas.

The EEC directives relating to the permitted sound level and exhaust devices of motor vehicles (EEC/157/70; EEC/350/73; EEC/212/77) were implemented by Law 942 of 27 December 1973 and subsequent ministerial decrees of 5 August 1974, 26 August 1977 and 5 May 1979.

Similarly, the directive of 17 December 1975 relating to the noise of motorcycles was implemented by Ministerial Decree on 5 May 1979.

The Transport Minister has the option of making inspections to ensure that the silencing equipment of the vehicle is functioning correctly.

In the case of industrial vehicles, the Minister provides for checks once a year and not once every 5 years as is the case for other vehicles.

6.2.1.2 POWERS OF THE PUBLIC AUTHORITIES

According to the TU 393 of 1959 on road traffic:

the absence of an audible warning device or its non-conformity with the law is punishable with a fine of from 4000 to 10,000 lire (article 46, clause 3);

the absence of a silencer or its non-conformity with the requirements of the law is punishable with a fine of from 4000 to 10,000 lire (article 45, clauuse 4);

failure to submit a vehicle for overhauling is punishable with a fine of from 4000 to 10,000 lire;

people causing a disturbance or driving with an altered silencer are subject to a fine of from 5000 to 20,000 lire;

Violation of the permitted use of an audible warning device is punishable with a fine of from 4000 to 10,000 lire.

The regulatory powers of the mayor are widely used in tourist resorts during the summer months with regard to restricting the use of particularly noisy vehicles (motorcycles, etc.), especially at night. The communal authority also makes use of its powers to control the use of vehicles to limit the amount of noise produced by traffic.

6.2.2 Aircraft

6.2.2.1 CONTROLS OVER DESIGN, MAINTENANCE AND USE

In terms of aircraft noise, Italy conforms to the US Federal Aviation Regulation no. 36 of 1969: it is on this that RAI (Italian Aeronautical Register) bases its control over the design and construction of aeroengines.

The ICAO Convention of 2 April 1971 (in force in Italy since 6 January 1972) and the 1972 supplement provide the parameters to which the concession of navigability of aircraft are subject, in relation to the prevention of dangerous or disturbing aircraft noise. On the basis of this Convention, RAI must notify ICAO of any divergence from the international requirements on aircraft engine noise regulations in the approval of engines set by the technical regulations of the RAI.

The control of air traffic, including the requirements relating to not over-flying urban areas and the restrictions on night flying (which are, however, without much importance), is carried out by the Flight Assistance Inspectorate (ITAV) and by the General Direction for Civil Aviation at the Ministry of Transport.

6.2.2.2 POWERS OF THE PUBLIC AUTHORITIES

Powers regarding approval of aeroengines and their conformity with relative obligations and requirements are identical to those described in Chapter 2 (2.3). In terms of noise pollution from aircraft, there was a notorious application of article 659 of the Penal Code: the Magistrate of Monza condemned the Director General for Civil Aviation for noise pollution in the area surrounding the Milan airport at Linate (this was an application of clause 2 of the provision, combined, of course, with a violation of the State laws since the local authorities have no powers on this subject).

7
Nuclear Energy

7.1 INTRODUCTION

7.1.1 Legislation on nuclear matters

Nuclear legislation in Italy, as elsewhere, is based on criteria of flexibility. The basic principles and framework regulations remain constant and adaptations are made as time goes by, in line with the progress of knowledge and scientific, technical and industrial development, by means of further legal provisions at the same or at a lower level.

There are in existence 'technical' safety and protection regulations with regard to ionising radiation and nuclear energy. These regulations, issued by organisations and committees, concern the various aspects of industrial and professional activity and are mainly the work of UNI (National Unification Body of which the Nuclear Energy Commission—UNICEN—is part) and of CEI (Italian Electrotechnical Committee).

The regulations in question are not mandatory but are indicative of good technique in use, and their non-observance could lead to culpable responsibility.

Nuclear legislation centres round a few laws specifically drafted for the control of nuclear activities and the use of radioactive materials. Other laws are used in a supplementary role. Among the latter, the following come to mind: Law 1636 of 3 December 1922 (and RD 2440 of 31 October 1923) on the research and utilisation of radioactive substances; the TU of the health laws (RD 1265 of 27 July 1934 amended by Law 422 of 1942 and Law 1528 of 1942); Law 1103 of 4 August 1965 on the operation of medical radiological techniques; as well as the various laws on the safety of the work environment (in addition to Law 653 of

1934 and Law 860 of 1950 on work by women and children; and to DPRs 547 of 1955 and 303 of 1956 on health at work and the prevention of accidents; above all Law 300 of 1970, the 'Workers' Statute' is of particular significance, since under article 9 it gives the workforce a supervisory power and the right to take the initiative concerning safety at work).

However, the specific legislation on installations is based on a few fundamental laws. Law 933 of 11 August 1960 set up CNEN (National Committee for Nuclear Energy) and at the same time abolished the National Committee for Nuclear Research which had been instituted by Prime Ministerial Decree of 26 June 1952 and amended by similar decree of 24 August 1956.

Law 933 was revoked by Law 1240 of 15 December 1971 with the exception of articles 12 to 18. Law 1240 of 1971 substantially confirmed the powers of CNEN, whereas its organisations were almost entirely replaced. The law in question also contains provisions concerning the National Institute of Nuclear Physics (INFN) as a public legal entity with its own independent budget.

With Law 84 of 5 March 1982, CNEN changed its name to the 'National Committee for the Research and Development of Nuclear Energy and Alternative Energies (ENEA)'. Article 1 further specifies that in all provisions of laws or regulations in force the words 'National Committee for the Research and Development of Nuclear Energy and Alternative Energies (ENEA)' should be substituted for 'National Committee for Nuclear Energy (CNEN)'. In this chapter we have followed this directive of Law 84 of 1982, substituting the initials ENEA for CNEN when considering also the provisions of law precedent in time to Law 84 of 1982. Law 84 of 1982 modifies Law 1240 of 15 December 1971. According to this last law, CNEN had the task, amongst other things, to carry out and promote studies, and, in collaboration with the national industries specialised in the field, to promote the planning, construction and development of experimental and pilot prototypes, including those related to nuclear fuel, for the ends of the peaceful uses of nuclear energy. ENEA, instead, has the task of promoting the development and the qualification of the national industry in the respect of health and the environment. To this end it carries out and promotes studies, research, etc., pertaining to energy technologies falling within its competences and to energy savings, and further, as did CNEN, it takes care of the planning and the realisation of installations and component prototypes in collaboration with the industrial operators (cf. article 2 of both laws). In essence, with Law 84 of 1982 the tasks of CNEN are expanded and more detail is given to certain provisions of Law 1240 of 1971. The very name of the body as changed indicates that it also deals with alternative sources of energy, which, in article 1, are defined as being those taken from sources other than hydrocarbons. An interesting point of Law 84 of 1982 is that ENEA is to promote, carry out and coordinate studies, research and experiments on the environmental and health consequences of the exploitation and use of the sources of energy for the people in charge or for the population. This seems to indicate a greater awareness of the importance of the study of environmental impact, even if it does not yet indicate the development of a system or of a systematic application of an environmental impact assessment procedure in relation to the measures taken and the policy adopted.

ENEA (article 1 of Law 1240 of 1971) is a juridical person under the supervision of the Ministry of Industry, Commerce and Crafts and operates under the direction of CIPE. It is composed as follows (article 4):

The President, nominated by the President of the Republic on the suggestion of the Minister of Industry. His term of office is 5 years and he can only be re-elected for one further term (i.e. total possible term of 10 years).

The Board of Directors consisting of the President, eight members (of whom five are experts in nuclear technology, two are experts in company management and one is an expert in industrial technology), two experts designated respectively by the Minister of the Budget and Economic Planning and the Minister for the Coordination of Scientific Research; the Director General of Energy Sources and Industries from the Ministry of Industry; and three employees of the organisation.

The Executive Council consisting of the President as chairman and four members elected from the Administrative Council.

The College of Auditors consisting of three regular members and three substitute members with a term of office of five years.

Law 1860 of 31 December 1962 on the peaceful use of nuclear energy constituted the first important provision, with particular reference to industrial and scientific installations. This provision was necessitated by, amongst other things, the urgent need for legislation following the constitution of the European Community for Atomic Energy (EURATOM). The law was modified and integrated first by DPR 1704 of 30 December 1965, to conform with the directives issued on the basis of the EURATOM Treaty. Subsequently, Law 1008 of 19 December 1969 intervened with the intention of overcoming the difficulties encountered in the application of the declaration and authorisation procedures required for the storage, sale and transport of radioactive substances. DPR 185 of 13 February 1964 is extremely important. It was issued as a result of the powers delegated to the Government to pass laws on certain matters under Law 1860 of 1962, and contained a comprehensive regulation for the protection of the population from ionising radiation as well as measures for plant safety. There is a specific law with regard to the siting of nuclear installations: Law 393 of 2 August 1975 (concerning nuclear power stations and the production and use of electricity), which amends and integrates the regulations concerning nuclear matters already contained in Law 880 of 1973 (on the siting of installations for the production of electricity in general). Finally, there is DPR 519 of 10 May 1975 which integrates the regulations in Law 1860 of 1962 regarding the civil liability resulting from the peaceful use of nuclear energy: this DPR which was issued under the power delegated to the Government by Law 109 of 1974 ratifying the Conventions of Paris (1960) and Brussels (1963) on the use of nuclear energy, brings the regulations on the subject into line with the international directives.

7.1.2 Responsible authorities

The responsible authority at central level is the Ministry of Industry which operates in conjunction with other State authorities depending on the different uses of radioactive substances (Ministries of Public Works, Education, Health, etc.).

At peripheral level the responsible authorities are the Provincial Medical Officer, the Prefect and, in a limited way, the Labour Inspectorate and the Harbour Master.

The regions only carry out administrative functions in terms of the storage, sale or depositing of radioactive substances and control of environmental radioactivity on the basis of DPR 4 of 1972 (see 1.1.3): see articles 1, 6 and 13.

In addition, article 10 of DPR 185 makes provision for an Interministerial Council for Coordination and Consultation presided over by the Director General of Energy and Industry, and consisting of representatives from various ministries. The Council, whose members are in office for 4 years, carries out a coordinating function, on an administrative level, of the activities of the various Administrations; in addition, it expresses opinions on draft laws on the subject and on requests for revision of the basic provisions, according to article 2 of the Treaty of the European Community for Atomic Energy (EURATOM).

ENEA has already been mentioned under 7.1.1. It should be added that DPR 185 of 1964 has instituted a technical commission within ENEA for nuclear safety and health protection from ionising radiations.

The National Institute for Nuclear Physics (INFN) should also be mentioned. It is an organisation of a public legal status, with its own budget and important powers of experimentation and research: its forward programmes are approved by CIPE to which they are forwarded by the Ministry of Education.

As can be seen, CIPE also plays an important part in the planning of nuclear activities.

7.2 NUCLEAR INSTALLATIONS

This subject is largely controlled by DPR 185 of 13 February 1964 which constitutes the basic legislative text in terms of health protection, and by the subsequent implementing DPR.

This regulation also applies to apparatus generating ionising radiations (defined by DPR 1428 of 1968) which can result in risks of radiation.

7.2.1 Siting of installations

A general control over the siting of installations takes place when authorisation is granted: Law 1860 of 1962 requires installations to be authorised by decree from the Ministry of Industry after having heard ENEA.

The documentation to be supplied by the applicant should include an overall design and topographic plan of the installation together with an indication of the physical, meteorological, demographic and ecological characteristics of the zone involved, as well as a study of the methods for disposing of the radioactive wastes.

In fact, the siting of the installation constitutes one of the assessment parameters on the basis of which ENEA compiles the technical report required under article 39 of DPR 185. The Ministers of the Interior, Labour, Health and any others involved are requested to express their opinions on this report.

On the receipt of the above-mentioned opinions and the final technical opinion from the appropriate technical commission, ENEA conveys its consultative opinion to the Minister of Industry and it is this consultative opinion which represents the support on which the ministerial decree granting authorisation for the project rests.

The local authorities also have powers regarding this procedure (under their town planning and health and hygiene powers: see RD 1265 of 27 July 1934, TU health laws).

Article 38 of DPR 185 of 1964 extends the procedure in question (and confirms the need for approval from the Minister of Industry after hearing ENEA) to include nuclear installations intended for electricity production, including those which are not subject to authorisation under Law 1860 of 1962.

With regard to the siting of installations producing electricity, a comprehensive control is in force resulting from two laws: Law 880 of 1973 concerning the 'siting, construction and management of new thermal installations for the production of electricity' (and which has given rise to disputes regarding its applicability to nuclear installations); and Law 393 of 1975 on the siting of nuclear power stations and on the production and use of electricity (specifically aimed at controlling the siting of nuclear power stations). The procedures and controls resulting from this legislation are examined below.

The first phase concerns the determination of areas from which a site can be chosen for the nuclear installation.

On the basis of forward programmes prepared by ENEL for the construction of nuclear power stations within the framework of the national energy plan, approved by CIPE (in agreement with the Inter-regional Consultative Commission with the previous opinion of ENEA), it is decided in which regions the power stations will be sited. In consultation with the communes involved, with the Minister for Health and ENEL, the regions must indicate within 150 days at least two areas on which ENEA has given a favourable opinion.

If the regions do not put forward two proposed sites, the areas are determined by law on proposal of the Minister of Industry together with the Minister of the Budget (article 2 of Law 393 of 1975).

The second phase is of a technical and preliminary nature involving the assessment of the pre-selected sites.

Within 12 months ENEL must present a comprehensive report with indications of the location and documentation on the technical and environmental characteristics. This report is sent to the Minister of Industry, to the regions and to ENEA. Other Administrations involved (including the Ministry for Cultural Affairs) are asked for their opinion which is taken to be positive if it is not received within 60 days. ENEA carries out a technical investigation regarding the siting of the power station and presents its final opinion within 8 months to the Minister of Industry and the regions involved (articles 3 and 4 of Law 393 of 1975).

In a third, final, phase of the definitive siting of the power station, the region, in agreement with the communes, chooses the site within 60 days. If, within that period, the region has not done so, a further intervention is provided for: the final choice is made by CIPE.

7.2.2 Design and construction

The operation of nuclear installations is, as already mentioned, subject to authorisation by the Minister of Industry (article 6 of Law 1860 of 1962).

DPR 185 of 1964 prescribes the requirements to be observed by the applicant for the project to be approved (article 37 ff.). The documentation produced by the applicant supplies ENEA with information to permit it to make a pronouncement regarding the technical characteristics of the installation, the safety and protective measures and methods of disposing of wastes. On the overall design, the technical commission of ENEA and any other administrations involved can put forward observations and comments which will be conveyed to the Minister by ENEA when an opinion is given on approval for the project. Once the approval or permit is obtained for the overall design, the applicant cannot, however, construct or put the installations into operation until ENEA has given its approval for the detailed design of those constituent parts of the installation which ENEA considers important for the purposes of nuclear safety and health protection.

There are special controls for installations intended for industrial purposes or scientific research which could result in danger to the external environment or in which the total radiation is particularly high, or installations containing notably powerful apparatus generating ionising radiations. These must be authorised by the Minister of Industry in agreement with the Minister of the Interior, Labour and Health, having heard ENEA (article 55 of DPR 185).

The authorisation decree can contain all the requirements for the construction and operation of these installations, under the supervision of ENEA. The installations to which these controls apply are specified under DM of 4 January 1977. They include, for example, irradiation installations which can be used for a variety of medical treatments, particle accelerators, etc.

The construction is carried out under the technical control of ENEA which ensures that the construction corresponds to the approved plans (article 42 of DPR 185 of 1964). Once constructed, the installations are subjected to a 'test' using the methods laid down in articles 44, 45 and 46 of DPR 185 of 1964.

There are two types of test:

(i) *non-nuclear*, preceding the loading of the fuel or, in the case of installations for treating spent nuclear fuels, before they are introduced.

(ii) *nuclear*. ENEA can have its own inspectors present at both the first and second tests and minutes are compiled giving the results. Once these tests are complete, the installation is authorised to operate by decree from the Minister of Industry, having heard ENEA. The operating licence contains indications showing the maximum power capacity that can be achieved by the nuclear installation. This operating licence is in fact granted for successive operating phases depending on the positive outcome of successive groups of nuclear tests (article 9 of Law 1860 of 1962 and article 51 of DPR 185 of 1964): when giving its opinion to the Minister of Industry on whether or not to grant an operating licence, ENEA can also prescribe the observance of limits and conditions.

7.2.3 Maintenance and functioning

DPR 1450 of 30 December 1970, issued in application of article 9, clause 2, of Law 1860 of 1962, establishes the requirements and relative procedures for obtaining the qualifications required for the management and operation of nuclear installations.

The above decree, confirming that the personnel in charge of the technical operation of the installations must be 'suitably qualified', underlines the difference between 'management' and 'operation', the former referring to all the decision-making, organisation and coordination, the latter referring to the simple operation (article 3).

An *aptitude certificate* which is valid for 3 years and is renewable, is required to manage a nuclear installation. These certificates are of 'first class' or 'second class' aptitude (article 5).

The first class aptitude certificate is also valid for the management of an installation which requires a second class aptitude certificate. To qualify for a first class aptitude certificate, a degree in engineering, physics or chemistry is necessary; for the second class aptitude certificate a secondary school diploma is required (as a technical, electronics, physical, chemical or nuclear expert etc.). The second class aptitude certificate can also be granted to applicants who have attended the first two years of the degree courses specified above. In both cases it is necessary for the applicant to have spent at least 30 days at the installation to which his application refers (article 10). Applicants for the aptitude certificate must be between 21 and 45 years of age. The certificate is granted by the Inspectorate of Labour (article 12) once the appropriate commissions nominated by ENEA have carried out physical, psychological and professional checks on the applicants. The same Inspectorate also grants renewal certificates on application within 30 days from the date of expiry, providing the applicant is under 65 years of age (article 13).

For the 'operation' of a nuclear installation a suitable qualifying 'licence' is necessary, also valid for 3 years and renewable (up to the age of 65 years) and subject to the possession on the part of the applicant of a specific mental–physical aptitude and practical capacity to be verified by the commissions previously mentioned.

The applicant, who must be between 21 and 45 years old, must undergo a period of apprenticeship (article 20) of at least 60 working days, using a special personal employment card as proof of this apprenticeship.

This qualifying licence to operate nuclear installations can also be of two types: first grade (for supervisors, requiring a high school diploma or the first two years of a degree course in physics, chemistry or engineering); second grade (for operators, requiring the completion of compulsory schooling only).

DM of 1 March 1973 has provided examples for aptitude certificates, qualifying licences and personal employment cards required under DPR 1450, to be granted to those applicants who have been considered suitable for the management or technical operation of the above-mentioned installations.

The installations controlled by DPR 185 of 1964 are subject to operating regulations approved by ENEA having heard the technical commission.

The operating manual which the authorisation holder must exhibit for the purposes of the nuclear tests must also include as an annex an instruction manual for exceptional situations which might result in a nuclear emergency or in the expectancy of such an emergency.

For safety purposes, a specified number of qualified personnel must be present at the installation at all times. This number and their qualifications are specified individually for each installation by decree from the Minister of Industry. The working shifts for these personnel are established in the service order affixed to the place of work by the authorisation holder for the operation of the installation.

Some installations specified under article 8 of DPR 185 (*a*), (*b*), (*c*), (*d*), (*e*) and (*f*) require the constitution of a Group of delegates for installation safety, the composition of which must be approved by ENEA. This

Group has consultative powers concerning plans to modify the installation or to alter the operating procedures, programmes of extraordinary experiments, tests and operations, and proposals and recommendations on safety and protection. In addition, it must formulate a plan for internal emergencies and any subsequent modifications, and also assist the Director or Head of the installation in the adoption of measures aimed at dealing with events or abnormalities which could lead to danger to the public or to property.

A nuclear expert designated by ENEA attends meetings of the Group when they are concerned with the adoption of such measures (attendance of other meetings by the nuclear expert is optional).

In the previous section we discussed licensing procedures whereby the Minister of Industry authorises the operation of the installation following the positive outcome of the various tests, conditional on the observance of the permitted limits and conditions laid down by ENEA.

Here it is sufficient to add that ENEA ensures that all these requirements and conditions are adhered to. In addition, DPR 185 of 1964 states that the operator must keep all the appropriate operating records for all phases up to date (article 51).

7.2.4 Radioactive wastes

In this section, all the definitions and regulations concerning the collection, transport, treatment and storage of radioactive wastes will be dealt with. Article 104 of DPR 185 of 1964 states that

> 'whoever produces, treats, handles, uses, trades in, holds or stores natural or manufactured radioactive substances, must take the necessary measures to make sure that the collection, transport or disposal of solid, liquid and gaseous radioactive wastes are carried out in such a way as to ensure that no direct or indirect danger or damage is caused to single individuals and the population.'

Wastes can therefore be solid, liquid or gaseous.

(i) *Liquid wastes*: this normally applies to all liquid effluents containing radioactive nuclides or mixtures arising from the processing of radioactive materials. Working areas are classified into two types of zone for the purposes of ways of discharging liquid radioactive wastes and of the subsequent provisions to be adopted for their removal: *zones for contained discharge* and *zones for free discharge*. The zone for contained discharge is the area in which more than

0.1 μCi/day of highly radioactive substances and more than 10 μCi/day of medium to low level radioactive substances can be handled. In the zones for free discharge, the discharges are linked to the dirty water system or to the clean water system of the Centre or of the Laboratory. In order to avoid the daily handling of a greater quantity than envisaged in the zones for free discharge, the director of the laboratory or installation must record and declare the amount discharged at quarterly intervals to the Physical Health Service. In the zones for contained discharge the liquid wastes are collected in suitable tanks or reservoirs and emptied at the request of the installation or plant by the Physical Health Service.

(ii) *Solid wastes:* each zone where there is the possibility of this type of waste must be equipped with suitable 'waste bins', the activity of which must not exceed 50mR/h.

(iii) *Gaseous wastes:* these include gases and fumes arising from the processing of radioactive materials. The concentrations of gases in areas where people are located are fixed by article 87 of DPR 185 of 1964. Installations producing gaseous emissions must avoid all discharges into the atmosphere by adopting appropriate technical measures and suitable filters. They must also be equipped with apparatus for the control and sampling of gases.

A preliminary study of the disposal of the radioactive waste must be presented with the application for an authorisation under article 6 of Law 1860 of 1962.

The disposal of wastes of all types is authorised by the Provincial Medical Officer, having heard the Provincial Commission, for medical activities (article 89), and by the Prefect for industrial and scientific activities. This authorisation is not required for wastes with a very low radioactive content defined by DM of 14 July 1970, which establishes conditions for certain exemptions.

If, during the collection, transportation or disposal of wastes, an event occurs which might lead to significant contamination of water, air or soil, the persons carrying out the operation in question are obliged to notify the Prefect and the Provincial Medical Officer, and also the Head of the Maritime Department if a port or other area of the public seaboard is involved.

7.2.5 Obligations on the part of the installation operator

We have already discussed the requirements imposed on the operator of an installation relating to operational control, the manual of instructions for exceptional situations, essential personnel, operating records, the obligation to notify the Prefect or Medical Officer in cases in which the waste collection or disposal operations give rise to dangerous events. In addition to these, there are certain actions which he must carry out for reasons of safety.

In the first place, the operator must provide equipment for permanent monitoring of the level of radioactivity in the atmosphere, water, soil and foods, and he must provide for the analysis (article 57 of DPR 185 of 1964).

A further series of important obligations is foreseen for the protection of workers inside the installation: the employers and directors are required to take the necessary precautions for the safety of the employees and also to inform them of any risks, the methods of operation and standards of behaviour. In addition, they are obliged to ensure that the controlled areas are clearly delimited and marked with appropriate signs.

There are similar obligations to supply information to self-employed workers undertaking tasks inside the installation. Specific duties also relate to the limits of the doses and concentrations of irradiation to which workers or employees are subjected.

More generally, article 103 of DPR 185 of 1964 states that if during operations involving radioactive substances environmental contamination occurs, the employers, directors, people in charge or anyone who for reasons of work is responsible for eliminating the danger of further contamination or damage to individuals, must immediately notify the Provincial Medical Officer of the possible danger of contamination spreading to individuals or to air, water or soil in non-controlled areas.

For further obligations linked to the holding of radioactive substances, see 7.3.5.

7.2.6 Powers of the public authorities

We shall discuss the authoritative powers for enforcing the law referring to the powers of authorisation and control mentioned above.

The law states that the Minister of Industry can suspend or revoke the operating authorisation of the nuclear installation in the event of non-observance of the requirements for operation laid down at the time of authorisation or granting of approval. These measures can only be taken after the operator has been informed of the non-observance in question, regarding which he then has 30 days in which to provide a justification. Once this term has expired the Minister must allow the operator a further term in which to comply with the requirements before taking further action. Suspension of the authorisation or approval can be applied in urgent cases for the purposes of nuclear safety or health protection, but can last no longer than six months.

Revocation of the authorisation or approval by the Minister of Industry in agreement with the Ministers of the Interior, Labour and Health, and having heard ENEA, instead, follows repeated or serious non-observance of the requirements.

The authorisation or approval holder is liable to prosecution for contraventions in the event of his incorrectly carrying out the detailed plans for the installation, violating the operating regulations, violating the requirements contained in the authorisation, the approval or operating licence, etc.

There are nuclear emergency plans in relation to the safety of the nuclear installation and protection from radiation (articles 112–122 of DPR 185 of 1964 and circular 70 from the Minister of the Interior, 8 August 1973). The emergency plans consist of a series of coordinated measures to be adopted by the responsible authorities to ensure the protection of the population and property in the event of nuclear incidents. The plans must be reviewed every 6 months and updated every 2 years. The plan is compiled by a committee instituted at the Prefecture and composed of State and peripheral regional authorities such as the Chief of Police, the Head of the Fire Service, the Commander of the Carabinieri, a representative of Comiliter (Territorial Military Command), an official of the Ministry of Transport, the Provincial Medical Officer, the Provincial Veterinary Surgeon, etc.

The compilation of an external interprovincial emergency plan is also required for cases involving the foreseeable spread of nuclear danger to more than one province.

For the purposes of implementing the plan, the Prefect is given wider powers of intervention (article 122 of DPR 185 of 1964) including:

> the power to determine by ordinance the danger area and indicate controls by ordinance relating to the access and movement of people and goods;
>
> the power to supervise all intervention and relief services;
>
> the power to adopt all necessary measures.

If a director responsible for the installation fails to notify the authorities of an external nuclear emergency, he is punishable by imprisonment for a period of from 6 months to 1 year or by a fine of from 1,000,000 to 10,000,000 lire.

Anyone disobeying the orders of the Prefect during an external nuclear emergency is punishable by up to 6 months imprisonment or a fine of from 50,000 to 1,000,000 lire if the act does not constitute a more serious offence.

7.2.7 Rights of the individual

See 1.4.

With regard to the nuclear problem in particular, civil liability for damage resulting from the peaceful use of nuclear energy should be examined.

This responsibility is controlled by articles 15–24 of the basic nuclear law (Law 1860 of 31 December 1962, as modified by DPR 519 of 10 May 1975), following the ratification by Italy of the Conventions of Paris (1960) and Brussels (1963). Article 4 of Law 109 of 12 February 1974, implementing the two Conventions in Italy, delegated to the Government the power to issue within 1 year the necessary regulations adapting the legislation to the international directives. The basic principles relating to the responsibility for the peaceful use of nuclear energy can be summarised as follows:

(i) The law does not refer to all the risks that can involve people carrying out activities in this sector, but only those of exceptional gravity. The general legal rules concerning liability and insurance also apply.

In addition, the absorption of radioactive substances can produce damage which may be revealed a long time after the accident. This makes it advisable to fix a time limit for compensation. The limited area of applicability of special nuclear liability is clearly shown in article 15, clause 1, which holds the operator responsible only for damage caused by the nuclear incident or connected with it.

(ii) The operator of a nuclear installation is also responsible even if he can be shown not to have infringed the law and to have taken all suitable measures to avoid the damage; this does not apply only to nuclear incidents resulting from armed conflict or natural disasters of exceptional gravity (article 15, clause 4) (such events interrupt the chain of causality). Special rules are in force in the event of a nuclear incident concerning goods in transit (article 16).

The principle of concentrating responsibility on the operator of the nuclear installation is adopted as it facilitates the task of identifying the person to whom to apply for compensation. The operator can only take action against the physical person who intentionally caused the damage (in some cases in the measure predetermined in the contract).

(iii) The responsibility is not only restricted with regard to time but also to extent. The maximum indemnity limit is fixed at 7500 million lire.

(iv) The operator must have insurance to cover risks involving third parties and if, as a result of a nuclear incident, the validity of this insurance is reduced, it must be reconstituted by the operator to the extent and limits fixed by the Minister of Industry, otherwise the authorisation is automatically revoked.

(v) The law provides for State intervention (up to 43,750 million lire) in the event of a nuclear incident of catastrophic proportions, or which exceeds the insurance cover.

(vi) The sums provided by insurance or other guarantees can only be used to compensate damage caused by nuclear incidents. By law, they cannot be sequestered or distrained (article 21).

7.3 RADIOACTIVE SUBSTANCES

7.3.1 Storage and use

Law 1860 of 1962 states that whoever comes into possession of special fissionable materials in any quantity, or holds radioactive materials in a sufficient quantity so as to result in a total radioactivity exceeding one-tenth of a curie, must notify the Ministry of Industry within 5 days.

When the radioactive materials are in the possession of doctors or health organisations, notification must also be made to the Ministry of Health; in the event of these materials being held by university institutes for teaching or research purposes, notification must also be made to the Ministry of Education. This notification must be repeated annually before 31 December.

NUCLEAR ENERGY

Rules have been introduced by Law 1008 of 19 December 1969 and DM of 15 December 1970 which make exceptions to Law 1860 of 1962 with regard to holding (where the quantity of material held is small).

On the basis of article 1 of the DM mentioned, the storage of the following materials does not have to be declared.

(i) Substances in the form of metals, alloys, chemical compounds, mixtures, solutions and gas in which the natural or impoverished uranium or thorium content does not exceed 10 kg in total, or in which the concentration in weight of natural or impoverished uranium or thorium does not exceed 0.05% in total, when the total weight limit of 10 kg is exceeded (according to amendments contained in DM of 7 March 1973).

(ii) Rare metals, their compounds, mixtures and products which contain not more than 0.25 % in total weight of natural uranium and thorium

(iii) Minerals which do not contain more than 10 kg natural uranium or thorium in total.

(iv) Thorium contained in the following products: (a) mantles for gas lamps; (b) vacuum tubes; (c) electrodes for soldering; (d) electric light bulbs containing not more than 50 mg of thorium; (e) germicidal lamps and artificial sun ray lamps (not more than 2 g).

(v) Natural or impoverished uranium or thorium contained in the following finished products: (a) crockery and other ceramic articles (the varnish of which does not contain more than 20%); (b) vitreous products—vitreous enamel, vitreous or porous varnish which does not contain more than 10%; (c) photographic film, negatives and prints.

(vi) Any finished product or part of a finished product containing alloys or dispersions of tungsten–thorium or magnesium–thorium, provided that the thorium content does not exceed 4% of the weight.

(vii) Natural or impoverished uranium contained in the counterweights for aircraft.

(viii) Thorium contained in optical lenses (not more than 30%).

(ix) Thorium contained in any finished part of aeroengines made of alloy or dispersions of nickel–thorium (not more than 4%).

Trading in substances requiring notification is subject to authorisation by the Minister of Industry and Commerce (to be granted within 30 days) unless the European Community for Atomic Energy has the right of option under article 57 of the EURATOM Treaty.

Similar provisions are in force concerning import and export. The State, however, has the right of option on raw materials.

In addition to the declaration obligation, holders of these materials must also keep an accurate record in a special register under the requirements of DM of 27 July 1966 and DM of 19 July 1967. The decrees were issued on the advice of CNEN and they establish the methods for making declarations and for the keeping of records required under article 30 of DPR 185 of 1964. Anyone keeping irregular records is punishable with a fine (from 500,000 to 5,000,000 lire if the violation concerns special fissionable materials or raw materials; from 50,000 to 500,000 lire if the violation concerns radioactive materials or minerals: article 123 of DPR 185 of 1964).

For the purposes of trading in these materials, DPR 185 classifies operations into two types:

Type A: operations involving danger predominantly limited to the environment of the premises in which the trade is carried out;

Type B: operations also involving danger to the environment outside the premises in which the trade is carried out.

In addition to the requirements under Law 1860 of 31 December 1962, Type A operations must obtain a permit authorising the use of the premises, granted by the Prefect, after having heard the Provincial Medical Officer, the Inspectorate of Labour and the Head of the Fire Service (article 33 of DPR 185 of 1964).

A permit must be obtained from the Minister of Industry (in agreement with the Ministers of Labour, Health and the Interior: article 34 of DPR 185 of 1964) to carry out trading operations belonging to category B.

DM of 26 October 1966 establishes the provisions regarding the procedures for granting of permits. At present there are no similar provisions concerning the procedure for release of permits required from the Prefect for Type A trading operations.

The use of radioactive isotopes, determined by ministerial decree from the Minister of Industry by weight and radioactivity levels, is also subject to authorisation.

Article 13 of Law 1860 of 1962, subsequently modified by article 3 of DPR 1704 of 30 December 1965, assigns responsibility for granting authorisation: to the Minister of Industry together with the Minister of Labour for industrial uses; to the Ministers of Labour and Agriculture for agricultural uses; to the Ministers of Labour and Education for teaching uses; to the Ministers of Labour and Health for diagnostic and therapeutic uses.

The procedures for granting the authorisation are established by DM of 1 March 1974: the presentation of exhaustive documentation on the use of the isotopes, the maximum quantities used, the premises, etc. is required. This documentation is then submitted to ENEA for a technical opinion and to the Ministry of Labour or other interested ministries for an administrative opinion.

The authorisation is valid for 5 years and is renewable.

7.3.2 Packaging and transport

Firstly, on an international level, the 'IAEA Regulation on the transport of radioactive materials' should be examined. This Regulation was published for the first time in 1961 in Safety Series no. 6 and it has been revised twice, in 1964 and 1967. Since 1969 it has been adopted by nearly all the international transport organisations and numerous member States which used it as a basis for their own regulations on the subject. Study groups met in 1970 and 1971 and a final revised text was approved by the Governing Council in September 1972. The guiding principles used in the revisions were the maintenance of safety at the high standard provided for in the Regulation, the need to take into account scientific progress, and the opportuneness of encapsulating the Regulation within a fixed and durable framework. It is also envisaged that a study group will revise the Regulation every 10 years, but every 5 years the member States will be questioned on the need for anticipated revisions. According to article 102, the present regulation applies to land, sea and air transport, including personal transport, of radioactive materials which do not constitute an integral part of the means of transport, and according to article 104, it does not apply to the inside of installations in which the radioactive materials are produced, used or stored, excepting the act of storage during transport and for which other appropriate safety regulations apply.

The transport packages and containers must belong to one of the following three categories:

Category I	White
Category II	Yellow
Category III	Yellow

According to article 5 of Law 1860 of 31 December 1962, amended and integrated by article 2 of DPR 1704 of 30 December 1965, the transport of special fissionable materials in any quantity (but see Law 1008 of 1969 and DM of 15 December 1970) and of radioactive materials with a total radioactive quantity and weight exceeding the values contained in article 1 of DPR 185 of 13 February 1964, must be authorised by the Minister of Industry and Commerce by a special, specific decree issued together with the Minister of Transport and Civil Aviation or the Minister for the Mercantile Marine, depending on whether land, air or sea transport is involved.

According to article 1 of DM of 27 July 1966, ministerial authorisation is not necessary for the occasional transport of radioactive materials when the total quantity does not exceed:

(i) 10 millicuries per single nuclide of very high radiotoxicity included in Group I of the table annexed to DM of 27 July 1966, updated by DM of 19 July 1967 (for sealed sources of radium-226 (^{226}Ra), the limit is 300 millicuries);

(ii) 100 millicuries per single nuclide of high radiotoxicity included in Group II of the above mentioned table (for iodine-131 (^{131}I), the limit is 300 millicuries);

(iii) 1 curie per single nuclide of moderate radiotoxicity included in Group III of the table;

(iv) 10 curies per single nuclide of weak radiotoxicity, included in Group IV of the table.

Occasional sea transport of radioactive materials does not require authorisation when the total quantity of radioactivity does not exceed 2000 curies. In this case it is sufficient to make a declaration to the Prefect and to the Provincial Medical Officer of the provinces in which the transport commences and terminates at least 48 hours prior to commencement.

The transport of all special fissionable materials, regardless of the quantities involved, is always subject to authorisation from the Ministry of Industry and Commerce.

Finally, DM of 16 February 1976 (approval of the model guarantee certificate for the transport of nuclear materials) should be mentioned. This decree was issued in accordance with the requirements of article 2 of DPR 519 of 10 May 1975 amending articles 15–24 (nuclear civil liability) of Law 1860 of 31 December 1962 on the peaceful use of nuclear energy. Following this article 2, the new text of article 16, clause 3 of Law 1860 requires the operator in charge in conformity with this law to deliver to the transporter of the nuclear material a certificate provided by or on behalf of the insurer or other party providing the financial guarantee required under the same Law 1860 of 1962. This certificate must be established with a decree from the Minister of Industry in conformity with article 4 (c) of the Paris Convention (29 July 1960) on civil liability in the field of nuclear energy. The Convention was ratified and implemented in Italy by Law 109 of 12 February 1974, and the model certificate approved by DM of 16 February 1976 is the one recommended by the Nuclear Energy Agency in Paris (NEA).

7.3.3 Radioactive wastes

As already mentioned, under article 89 of DPR 185 of 1964 every province has a Provincial Commission presided over by the Provincial Health Officer and composed of two graduates in medicine (of whom one must be an expert in radiology), one physics graduate and a medical inspector of labour designated by the Inspectorate of Labour; it has a consultative function and remains in office for 3 years. This commission expresses opinions on authorisation applications for the following:

(i) the opening of medical institutes and consulting rooms;

(ii) the holding of natural or artificial radioactive substances;

(iii) the disposal of wastes.

With regard to the latter, the commission acts as a consultative organisation for the Provincial Medical Officer (who is responsible for authorising the discharges of radioactive slag from medical institutes, departments or surgeries) and the Prefect (who, having heard the Provincial Medical Officer, authorises the disposal of wastes from industrial or research activities).

7.3.4 Obligations of the holder of radioactive substances

In addition to the obligations of notification, record keeping and authorisation mentioned above, various protection obligations with regard to the workers and the population must be taken into consideration.

7.3.4.1 HEALTH PROTECTION FOR EMPLOYEES

This is supervised by the Ministry of Labour and Social Security through the Labour Inspectorate, as required under articles 58–87 of DPR 185 of 1964. Minors and pregnant women cannot be employed in occupations where workers are exposed to ionising radiations.

Among the obligations of employers are the following:

(i) the duty to provide workers with full information concerning specific risks;

(ii) to ensure physical supervision by means of authorised doctors and qualified experts;

(iii) to keep up-to-date minutes, personal records and health documentation (articles 74 and 81) which must be kept for at least 30 years after the cessation of employment;

(iv) to subject workers to periodic and preventive medical inspections. DPR 185 of 1964 also provides (under article 83) for a specific monitoring activity conducted by institutes authorised by the Ministry of Labour. A decree from the Minister of Labour in agreement with the Minister of Health and having heard ENEA fixes the maximum global, partial or accidental radiation doses and maximum permissible concentrations of radioactive nuclides in the air and drinking water, to be monitored by the employer.

Particular regulations are in force for mine workers working in the presence of radioactive materials or substances.

In addition to the subjects specified in DPR 185 of 1964 (articles 61 and 62), there is DM of 13 May 1978 ('Health safety and protection for workers in the mining industry against ionising radiation').

7.3.4.2 HEALTH PROTECTION OF THE POPULATION

This takes various forms but can be analysed as follows:

(i) declaration to the Provincial Medical Officer and to the Head of the Fire Service, of the loss or leakage of artificial or natural radioactive substances (article 94 of DPR 185 of 1964);

(ii) authorisation by the Provincial Medical Officer (after having heard the Provincial Commission, treated under article 89 of the DPR) for the opening of medical consulting rooms or surgeries etc. where radioactive substances are used, even if not regularly, for therapeutic or diagnostic purposes (article 96 of the DPR);

NUCLEAR ENERGY

(iii) prohibition on the professional use of X-ray diagnostics by medical graduates not equipped with a specialist diploma in radiology (article 97 of the DPR);

(iv)   registration in special registers of radioactive substances over certain limits administered to patients (article 98 of the DPR);

(v)   mention in death certificates of the nuclides administered;

(vi)   precautions for doctors and all medical personnel assisting at autopsies;

(vii) controls over contamination of the environment (obligation to take all necessary precautions to eliminate the danger of further contamination);

(viii) arrangement for suitable means of sampling and monitoring regarding the disposal of wastes.

In general, article 95 of DPR 185 of 1964 states that whoever produces, treats, handles, uses, trades in or holds radioactive substances must ensure that the population is not exposed to the risk of absorbing doses of radiation in excess of those limits fixed by the ministerial decree issued on 2 February 1971, and that the air and water are not contaminated in excess of the said limits (see 7.3.5).

7.3.5 Legal standards, objectives and guidelines relating to levels of radioactivity in the environment

Control of radioactivity in the environment and foods is the responsibility of the Ministry of Health.

ENEA (article 109 of DPR 185 of 1964) coordinates monitoring carried out by the administrations, institutes and organisations regarding the radioactivity of the atmosphere, water, land, foods and beverages; it promotes the installation of sampling stations and relative measuring of radioactivity; it passes information regarding the samples and measurements to the Commission of the European Community in accordance with article 36 of the European Treaty on Atomic Energy.

DM of 2 February 1971 fixed the values of maximum permitted doses and concentrations for the population as a whole and for single groups.

This decree implements article 111 of DPR 185 of 1964 and can be considered similar to the one issued in implementation of article 87 of the same DPR to set the maximum permitted doses and concentrations for workers exposed to ionising radiations.

The values in this decree also conform to those indicated in the EURATOM Directives on health protection.

The maximum permitted dose for the population as a whole from a genetic viewpoint is 5 rem per capita, accumulated up to 30 years of age (article 1).

The maximum doses are then fixed at different values for various parts of the body by article 2, depending on the three population groups, provided for under article 9 (*h*) of DPR 185 of 1964:

Group 1: people who for work reasons occasionally find themselves in the controlled zone but who are not considered to be people exposed for professional reasons.

Group 2: people who handle apparatus emitting ionising radiations.

Group 3: people who are habitually in the vicinity of the controlled zone and who for this reason may receive radiation doses in excess of the limit fixed for the population as a whole.

As we have seen, these limits, according to article 95 of DPR 185 of 1964, must not be exceeded even in the environment (water, air).

The Provincial Medical Officer can issue regulations to ensure that the duties to control these limits which fall upon those who use radioactive substances are complied with.

7.3.6 Rights of the individual

See 1.4 in general, as well as 7.2.7.

7.3.7 National Energy Plan

The approval by Parliament (22 October 1981) of the National Energy Plan should be noted. This Plan embraces the decade 1981–1990 and covers the different sources of energy (coal, nuclear, natural gas, hydro-electric and geo-thermal energy, solar, oil), analysing the present situation of each and making forecasts. Furthermore, it contains sections on Energy and Development, and on Security, Environment and Territory, and includes four appendices (A—Options for the installation of coal and nuclear thermoelectric power plants; B—Coordination programme for the construction of nuclear power plants; C—Infrastructures for the transportation of coal; D—Urgent actions for the putting into operation of the nuclear power plant of Caorso).

8
Control of Products

8.1 SYNTHETIC DETERGENTS

The question of the composition and quality of detergents is controlled by Law 125 of 3 March 1971, and its implementing regulation DPR 238 of 12 January 1974, and by provisions concerning the methods for applying the law (e.g. DM of 19 July 1974). There is also a European agreement restricting the use of certain detergents contained in washing and cleaning products (signed at Strasbourg on 16 September 1968), implemented in Italy by DPR 974 of 26 November 1976.

Law 125 of 1971 refers to synthetic detergents, stating in article 1 that for the purposes of protecting surface and groundwaters from pollution, they must be at least 80% biodegradable. At the same time the use of detergents of any quality is forbidden if they can cause damage to the health of man or animals. Article 2 of the law prohibits the production, storage, trade, import into the State or use by industrial or public installations of synthetic detergents which do not conform to the requirements. The violation of this prohibition is punishable as an offence by imprisonment for up to 6 months and by a fine of from 100,000 to 5,000,000 lire where the act committed does not constitute a more serious offence. The law refers to the legislation on the hygiene of production and sale of foods and beverages (Law 283 of 30 April 1962 and subsequent modifications) for everything to do with powers to monitor installations, manufacturing processes, storage, sale and consumption of detergents considered by the law, as well as the regulatory powers of the health authorities and the associated penal sanctions. In addition, article 5 of Law 125 of 1971 states that the Provincial Medical Officer can order the confiscation and, with permission from the judicial authority, the destruction of products not conforming to the requirements of the law (not only must they be at least 80% biodegradable, but they must also not be a danger to the health of man or animals).

Packets of detergents are required to show the exact percentage of biodegradability as well as all the elements necessary to identify the product and the producer, and the net weight or volume of contents.

The implementing regulation (approved by DPR 238 of 1974) contains further obligations and specifies the legal requirements: violation of the provisions of the regulation is punishable under article 6 of Law 125 of 1971 with a fine of from 50,000 to 500,000 lire where the act does not constitute a more serious offence. In the first place, biodegradability of at least 80% is required for all synthetic detergents as such, or synthetic detergents present in other detergents, and for mixtures of synthetic detergents as such or present in other detergents. The methods for determining the percentage of biodegradability are fixed by decree from the Ministry of Health. On this subject the DM of 19 July 1974 states that the determination of biodegradability must be carried out using a 'method of selection', employed for the control of all samples, and a 'method of confirmation' which is a second screening used only for the control of samples discarded by the 'method of selection', in order to confirm or cancel the results originally obtained. The DM describes in detail all the procedures for the preliminary treatment of the products to be examined and the execution of the various tests ('of selection' and 'of confirmation'), defining the values and the formulae which must be satisfied.

The Minister of Health is also competent for determining by Ministerial decree the method for defining the toxicity trial and the concentration values for synthetic detergents (the decree is issued in agreement with the Ministers of the Interior, Public Works and Industry and with the opinion of the Superior Health Council).

The regulation sets out precise requirements concerning the indications on packets of detergents and synthetic detergents (which, incidentally, must be written in Italian if the goods are destined for the home market). In particular, the following must be shown: category of synthetic detergent, percentage of biodegradability, name, company status, trademark, address of manufacturing and marketing company or companies, the location of the producing plant, instructions and conditions for use, net volume or weight of product.

According to the regulation, the Ministry of Health, its peripheral organs and the central and peripheral regional health organisations are the authorities in charge of monitoring with the assistance of the technical secretaries, health inspectors, Provincial and Communal Health Officers, Provincial Laboratories of Hygiene and Preventive Medicine and other public laboratories authorised by the authorities to undertake sampling and analysis operations. The Provincial Medical Officer (or other regional organisation responsible) coordinates monitoring to ensure uniformity of intervention and inspection criteria, sampling and notification in accordance with the directives of the Minister of Health.

The Minister of Health also has the task of authorising the operation of installations, laboratories manufacturing preparations and packaging and of wholesale storage of cleaning products and synthetic detergents. Applications for authorisation must contain (amongst other things) the following information: category of synthetic detergents used, and their degree of biodegradability and toxicity determined using the methods indicated in the regulation; health and hygiene precautions to be adopted during the manufacture, preparation, packaging or storing of the product for environmental protection purposes; documentation concerning the studies and research carried out to determine the effects on the health of humans and animals; analytical data on the individual components of a product and of the active ingredients for a thorough knowledge of the product itself.

The regulation describes in detail the sampling operations for controls and analyses. Records must also be kept on these operations. Four samples are required and must be obtained by dividing a single average sample: the first is left with the manufacturer, together with a copy of the Records; the others are sent to the Provincial Laboratory of Hygiene and Preventive Medicine (or other authorised laboratory) and used as follows: one for carrying out the analyses, one for any revised analysis required and the last one is held for any eventual assessment which may be ordered by the judicial authorities.

If the analyses show the product to vary from the legal requirements, it can be confiscated. If the revised analysis, which can be requested by the interested party, confirms the original result, the destruction of the product is ordered.

The provisions described above are substantially in line with Italy's commitments in the European agreement on the limited use of certain detergents in washing and cleaning products, adopted in Strasbourg on 16 September 1968 and implemented by DPR 974 of 26 November 1976. This agreement also requires the prohibition of detergents with a biodegradability of less than 80%, and of all products which are dangerous to humans or animals; all the signatory countries are required to take necessary monitoring and control measures.

In addition, the agreement in article 3 requires the parties to hold multilateral consultations at the Council of Europe every 5 years or more often on request, to control the implementation of the agreement and examine the necessity for reviewing or extending the provisions.

EEC Directives EEC/404/43 and EEC/405/73 on the method of controlling the biodegradability of anionic surface active agents have not as yet

CONTROL OF PRODUCTS

been explicitly implemented in Italy, even though they contain very similar provisions to those mentioned.

8.2 DISINFECTANTS, INSECTICIDES, PESTICIDES AND SIMILAR PRODUCTS

The legislation regarding the use of disinfectants, pesticides etc. (phytopharmaceuticals, agricultural chemicals, preservatives for food storage etc.) refers above all to their use in agriculture and dictates limits and prohibitions, but also specifies the methods for manufacture and use.

The production and sale of these products are controlled by DPR 1255 of 3 August 1968 to which are added DM of 19 July 1969, DM of 28 November 1970, DM of 28 December 1970, DM of 20 October 1971, DM of 29 July 1972 and DPR 424 of 9 May 1974 modifying or integrating its provisions. Companies authorised to produce and market these products are determined by Ministerial decree. Health protection agents permitted on the market are also determined by Ministerial decree.

The regions have various powers on the subject both regarding general responsibility in agricultural matters (article 117, Constitution) and as a result of the decrees transferring certain functions to the regions, by which they are empowered to authorise the storage and trade of phytopharmaceuticals and health protecting agents for agriculture and foodstuffs.

The Minister of Health has competence for granting authorisations for the operation of installations manufacturing the above mentioned products. The authorisation can be revoked if the manufacture or storage does not conform to the conditions and requirements prescribed for health protection or indicated by the law.

The Minister is also able, by Ministerial decree, to issue a list of authorised products, divided into four classes depending on their toxicity to humans and animals and according to the greater or lesser tolerable concentration in crops and foodstuffs (the maximum permitted concentration of active elements in drugs on sale is also rigorously determined by decree of the Minister of Health).

The use of certain substances is subject to strict limits or absolutely forbidden; there is a ban on the use of:

DDT (DM of 14 January 1970 for health protective agents; DM of 31 July 1973 for pesticides);

for basic active elements containing antibiotics, chemotherapeutics, or phenylmercuric acetate (DM of 10 August 1971);

health protective agents containing oil of creosote (DM of 7 October 1972);

all organic compounds of mercury for pesticides (DM of 7 October 1972);

health protective agents containing mixtures of meta- and paracresol (DM of 31 July 1973), quintozene (DM of 23 October 1973), monofluoracetamide (DM of 7 February 1974), aminotriazol (DM of 3 August 1974); hexochlorocycloesane (DM of 14 August 1974); for diallate, triallate and sulphate (DM of 8 July 1977).

Each new product, however, is subject to registration which can be conditional on certain requirements concerning its use and restrictions, or can be granted for a certain period only. The Consultative Commission at the Ministry of Health proposes registration, possibly after the Superior Health Institute has completed the necessary analyses, and the product is then authorised by decree from the Minister of Health.

The local health organs (Health Official and Provincial Medical Officer) are responsible for the authorisation and monitoring of the storage and sale of such products. In particular, the Health Official grants authorisations for the sale and an entitling certificate for the manager of the store. The authorisation can be revoked in the event of non-observance of the requirements of the law. The Provincial Medical Officer, together with the Health Official, carries out monitoring and control operations, also by means of inspections and sampling. In the event of danger to public health, they can prohibit the sale of determined products. In addition, the Provincial Medical Officer can decide on appeals made against a provision issued by the Health Official revoking authorisation for the sale of a product.

The law outlines punishments for violations of the registration and authorisation provisions: offenders can be imprisoned for up to 1 year and a fine of from 200,000 to 30,000,000 lire can be imposed.

Other requirements concern the packaging and labelling of products and compulsory records which must be kept in a suitable register of all incoming and outgoing operations regarding substances covered by the law (cf. DM 21 May 1981 on the classification and regulation of the packaging and labelling of dangerous substances, in implementation of the Directives issued by the Council and the Commission of the European Communities).

Directive EEC/324/75 relating to the harmonisation of the legislation of member States regarding aerosols is awaiting the promulgation of a law as delegated to the Government by Law 42 of 9 February 1982.

8.3 FOODS

The question of hygiene regulations on foods principally concerns problems which are not directly related to this book, but are related to the protection of public health and the consumer. Nevertheless, there are provisions aimed at avoiding certain forms of environmental deterioration or certain environmentally damaging activities which can result in direct harmful effects for man through substances intended for human consumption.

For example, the very harmful effect produced in humans by organic mercury accumulated in fish which are then consumed by man is a well known phenomenon. The Minister of Health has, through various ministerial decrees (DM of 14 July 1971, DM of 14 December 1971, DM of 21 December 1972, DM of 6 September 1973, DM of 29 March 1974) fixed the limits of permitted contamination for aquatic produce (0.7mg/kg), and the rules to be observed by fish importers: they must be in possession of a health certificate from the authority in the country of origin, which shows the concentration of mercury and the analyses carried out.

A problem only marginally connected with this study is the one concerning foods which have undergone special treatments with ionising radiation. Law 283 of 30 April 1962, containing hygiene requirements for the production and sale of foods and beverages, provides in article 7 a special regulation for the sale of foods which have been subjected to ionising radiation (see also DM of 30 August 1973 authorising, under certain packaging and labelling conditions, the storage and sale of products such as potatoes, onions and garlic treated with gamma rays to prevent germination and DM of 21 May 1981 mentioned above).

Another very controversial question (which has given rise to contradictory opinion even at official level) is the production of bioprotein, i.e. synthetic proteins used mainly in breeding-animal foodstuffs which could cause the concentration and accumulation of active toxic elements in the meat of animals. DM of 21 June 1977 was issued to control the measures for health, hygiene and environmental protection connected with the production of bioprotein, following a preliminary authorisation for the production of bioprotein from the Minister of Health (DM of 14 May 1972) and further controls and analyses requested by various parties on the basis of studies conducted by the Higher Institute of Health and on the opinion of the Higher Health Council. On this basis, the production of synthetic flour proteins can be authorised but is restricted to the experimental needs of industry, of a temporary nature and in limited quantities (the terms and rate of production of the installation are defined in the decree granting authorisation).

Manufacturing industries must carry out immunological and clinical tests on the groups and population exposed and, once the installation has commenced operations, they must carry out periodic checks on the workers and the population exposed. There are also requirements for protection of the external environment from contamination produced by liquid or gaseous emissions or by solid wastes: the minimum possible level of contamination must be pursued in addition by using controls instituted by the industry itself at its own expense. Monitoring is the task of the Regional Health Authority. The flour protein produced during the course of the authorised experiment must not be sold or used.

8.4 MINERAL OILS

The regulations on the import, production and storage of mineral oils are contained in Law 367 of 2 February 1934 (which converted RDL 1741 of 2 November 1933) and subsequent amendments, as well as the regulation approved by RD 1303 of 20 July 1934 and subsequent amendments. A licence is required for the importation of quantities of mineral oils which are not white oils nor petroleum coke. A permit is required (or authorisation for quantities of less than 5000 tonnes a year) for the processing, refining, distillation or mixing of petroleum products. Similarly, a permit is required (but unlike the provisions mentioned above which are the responsibility of the Minister of Industy, this permit is granted by the Prefect) for the installation and management of warehouses or distribution centres for petroleum products (with prior approval from the Chamber of Commerce, Head of the Fire Service, Technical Office for Taxation of Manufacturing). Naturally, the permit from the Ministry of Industry for the treatment of mineral oils, in the case of a new installation, must be accompanied by an authorisation from the authorities in charge of protecting the territory and public safety: if the installation is built on the shore, it comes under the authority of the Minister for the Mercantile Marine who is responsible for the protection of the seaboard and territorial waters; otherwise it is the mayor who makes the decision regarding the siting of the installation. There are special provisions in the form of ministerial provisions for some categories of mineral oils (most recently DM of 16 December 1977 controls the production, storage, circulation and use of benzole, toluol, xylol, paraffins, olefins and naphthanes).

Exemptions for businesses from the obligation to sell used oils to authorised regeneration plants or else to collect and regenerate them themselves for their own use have in fact been extended beyond the limits permitted under the law. In fact, this obligation, which should apply to public organisations, industrial installations, agricultural and transport businesses with a consumption of more than 50 kg of oil a month, and workshops and garages accumulating more than 40 kg of used oil a month, remains largely unobserved, even though it is backed up by sanctions (fines).

The provisions discussed above do not contain any specific requirements on the need for environmental protection, nor have they been used to control the polluting qualities and characteristics of mineral oils.

8.5 TOXIC GASES

The legislation on this subject centres round Law 2231 of 6 December 1925 and the TU of the laws on public safety as well as special regulation 147 of 9 January 1972, amended by DPR 854 of 10 June 1955.

Gaseous substances or dangerous toxic vapours are specified in a list of 12 categories, approved by decree from the Minister of Health (the last revision of the list was made by DM of 29 January 1978). The local health offices belonging to the regional organisation (Provincial Medical Officer) are responsible for health monitoring on the question, and, in particular, the use of such gases is subject to inspections and spot checks. In addition, the operation of manufacturing installations and the storage of toxic gases have to be authorised, with conditions and methods of use specified in the authorisation. The authorisation can be suspended as a result of non-compliance with the regulations on use, or with the general provisions contained in the authorisation itself, or imposed by law. It can be revoked for failure to take the necessary precautions imposed upon the planning and construction of fixed installations for manufacturing, processing or storage.

The use of toxic gases without authorisation is an offence punishable by imprisonment for up to 3 months and a fine of up to 80,000 lire.

The management of installations using toxic gases is restricted to holders of a special qualifying licence which is granted following an examination.

The use of toxic gases (when not used by industrial installations as part of the manufacturing process) is subject to a licence from the public safety authorities (the application must be presented within a period fixed by the law). The applicant must specify the measures adopted to diminish the risks and eliminate the toxic effects of the gases. In addition, the use of appropriate signs (or even barriers) is essential to denote the zone adjacent to the area in use. The residues from the used gases cannot be discharged into wells, watercourses, tanks, troughs etc. and must be neutralised or rendered harmless before being discharged into drains, canals or stretches of water.

Industries are allowed to store toxic gases only in the quantities necessary for the processes involved and any excess must be returned to the store on a daily basis. The removal and use of toxic gases must be listed in a register drawn up for this purpose which should be kept available for the public safety authorities.

The public safety authority has competence for granting a licence for the transport of toxic gases. Rigorous requirements are laid down concerning containers, safety, packaging and labelling and with regard to the times and methods of transportation by road (cf. DM 13 March 1981—Amendments to the Ministerial Decree of 30 October 1968 concerning specific regulations for the loading, transport by sea, unloading and transhipment of dangerous goods in packaging of the first class (explosives); DM 21 May 1981—Classification and regulation of the packaging and labelling of dangerous substances in implementation of the Directives issued by the Council and the Commission of the European Communities; and DPR 927 of 24 November 1981—Reception of the Directive of the Council of the European Communities No. 831/79 of 18 September 1979 with the Sixth Amendment to Directive EEC/548/67 concerning the classification, packaging and labelling of dangerous substances and preparations).

9
The Environmental Problem and the Italian Juridical Organisation

9.1 GENERAL REMARKS

Probably the most characteristic aspect of Italian juridical experience in environmental protection is the importance given to the search for the use, in new modes and forms, suited to the nature of the problem, of provisions formulated for totally different purposes and frequently remote in origin.

Not that there is a lack of more or less recent special legislation on the environment, but as reality must be acted upon through the mesh of an administration which is too segmented and disjointed to face the new problems of environmental protection (which require at the least more penetrating programming and coordination between administrations competent in different areas), the laws for environmental protection have severe limits on their perspective, each one facing a separate aspect of the problem which is a complex one. Besides, it is not the task of special legislation to create a coordinated framework for the public authorities. The laws on the protection of air, water and land should provide sufficiently technical and detailed regulatory instruments for an administration already able to operate effective territorial planning efficiently, thanks to a renewed structure of powers and responsibilities. The problem, therefore, is one of administrative organisation and, above all, of making the instruments existing within the regional organisation more effective, as well as the functions of minor organisations such as the communes, which are of great importance in this field.

It is therefore not surprising that the problem today centres around the attempt to use 'civil' remedies for the protection of individuals, by means of interpretations of traditional legislation which can be used to resolve conflicts between individuals exercising a certain activity and other individuals suffering from the damaging or annoying consequences of that activity. This is to say, not only is the problem of environmental protection first presented in Italy as a problem of protecting private interests (often only the interests of an individual) rather than a public interest, but the protection of the private interest has evolved as a result of the more or less successful search for an adequate interpretation of traditional legislation. These interpretations are often accepted by judges, with some degree of vacillation and hesitation, but also with growing technical perfection and firmness (it is sufficient to compare the early judgements of the 1960s pronounced by the leading judges, and particularly the magistrates, with the latest valuable judgements of the Court of Cassation: 9 March 1979, No. 1463 and 6 October 1979, No. 5172).

This circumstance brings to light a peculiarity in the 'Italian case' characterised by advanced legislation (particularly constitutional) for the protection of interests, including group interests, of the individual but at the same time by an uncohesive and episodic control of the environment.

In this context, protection of the environment through Civil Law, which has also developed as a 'substitute' for the lack of environmental planning (and therefore of a comprehensive administrative protection), presents an ever more striking duplication with regard to administrative protection, in the sense that any activity with repurcussions on the environment can be permitted from the administrative point of view, but nevertheless fall under censure from a civil judge, and vice versa. In fact, administrative regularity (e.g. a productive activity carried out once all the required authorisations and licences have been obtained) does not automatically preclude the injustice of the damage under article 2043 of the Civil Code (e.g. damage to the health of third parties arising from correctly authorised activities which nevertheless result in pollution).

This situation was inevitable. The incomplete nature of the preventive programme and of the overall management of the environment have necessitated the development of an instrument of protection for injured interests. This protection is of necessity intermittent in nature, linked to the reasoning behind each individual case and to the discretion of the individual judge involved. This results on the one hand in the laudable blocking of a loophole, and on the other introduces elements of anarchy into the case in question and uncertainty on the level of territorial policy in general. The future legislator who wishes to move towards a planned and comprehensive management policy for the territory and the environment must, therefore, also deal with the difficult problem of coordinating administrative protection with that through Civil Law.

The way in which the civil protection instruments, both direct and indirect, are interpreted with regard to environmental protection in Italy is summarised in the following paragraphs.

The interpretation of article 844 of the Civil Code, which controls the immissions of smoke, gas, vapour, noise and vibrations on to the property of others, is very symptomatic of the way in which regulatory instruments are applied in Italy. This regulation has been cited as an instrument of civil protection for individuals against polluting activities, despite the fact that its formulation in the Civil Code presently in force appears to confine it to the role of controlling relationships among neighbours and therefore as a limit to the use of the property imposed in the interest of the neighbour involved. It is also used as an instrument of reconciliation in conflicts arising from the different uses of buildings. We have in fact already seen the nature and extent of the many obstacles which hinder the use of the provisions controlling immissions for the protection not of property but of health, and above all of the health of individuals when they are not the owner of the property affected by the emissions. However, the opinion (VISENTINI) that the provision under article 844 of the Civil Code is in fact a special provision of responsibility for extra-contractual unlawful acts appears to be based on uncritical attempts to introduce certain constructions and schemes taken from other legislations (such as the French) into the Italian legislation; such attempts are incompatible with the logic, origin, function, content and systematic arrangement of article 844 of the Civil Code of 1942.

In a more correct perspective—which preserves the regulating function of article 844 of the Civil Code for the different uses of neighbouring properties and only aims at a review of the conflict of interests which is the basis of the regulation in a constitutional context—in addition, it has been maintained (SALV I) that it is necessary to overcome the position giving rise to this conflict of interests (which only took into account the economic importance to the national economy of the property and of the business involved) in order to introduce, when considering interests entrusted to the judge by article 844 of the Civil Code, the constitutional principles which impose limits of safety, freedom and human dignity on private economic initiative, and assign a 'social function' to property ownership. Despite the fact that it is undoubtedly true that protection of individuals and control of economic activities are closer together as a result of the constitution of 1948, the use of legal means is still problematic for the protection of the national heritage (such as property), and for the protection of personal values (such as health, particularly if it is understood as a psycho-physical balance, or as quality of life). Whereas the constitutional provisions of the 'function' and 'social use' of property and private initiative which are highly incisive as limits to entrepreneurial choice in contrast to these provisions (e.g. uneconomic and polluting activities) seem hardly in harmony with article 844 of the Civil Code; they do, however, raise the legitimisation of the interests of individuals (and/or businesses) in requesting judicial control over the violation of such limits, even where their 'interest in acting' (under article 100 of the Code of Civil Procedure) is not based on any injury to an 'individual right'.

However, it is widely believed that the regulation of immissions can provide only limited protection, exclusively individual and subject to a particular justifying situation which results from the legal relationship of an individual with the property which has been directly damaged by a polluting activity.

The statement of the problem in terms of compensation for extracontractual damage would appear more promising. Italian legislation concerning extracontractual civil responsibility is based on the atypical nature of the type of offence (see RODOTA, TRIMARCHI, SCOGNAMIGLIO, ALPA) and constitutes an open system under which any subjective legal situation can be protected where the injury can be considered to be 'unjust' damage under article 2043 of the Civil Code.

The control of extracontractual responsibility is one of the elements in Italian legislation with the greatest response to the new needs for protection resulting from the development of certain typical aspects of an advanced society, precisely because it is based on a somewhat elastic 'general clause'. For example, in terms of the responsibility of the manufacturer for defective and harmful consumer goods there are, in jurisprudence, solutions often based on reference to civil responsibility regulations (CARNEVALI, GHEDINI, BESSONE, ALPA, RUFFOLO, CASTRONOVO).

The use of this protective instrument in the field of damage to the environment can be facilitated by some of the developments in jurisprudence and practice relating to health protection. Thus, the consideration of the right to health as a protected right in itself and for itself, not only when injury leads to a loss of the earning capacity of the individual (BUSNELLI) as was the case in the past, has led to the possibility of obtaining compensation for damage to health even when it does not lead to a decrease in the working capacity of the injured party. It is therefore possible to distinguish between damage to health and damage to physical wellbeing (RUFFOLO), making a gradation of protection possible for the two types of damage. Finally, within this concept recognised by a more enlightened jurisprudence, it has been possible to arrive at a much wider notion of damage to health, which is understood as prejudice to the physical and psycho-physical balance and to a certain quality of life (see Cass. 6 October 1979, no. 5172, in *Foro it.* 1979, I). Based on the provisions of articles 2 and 32 of the Constitution, it has been recognised that the individual can claim a true subjective right, armed with penetrating powers of protection for his own physical and psychological wellbeing, which can be made even when that wellbeing is endangered by activities which harm assets external to the person yet conditioning his existence.

But in practice there are signs that even this advanced arrangement is being constitutionally superseded by way of the systematic interpretation of a right to the environment (PATTI) to be included among the rights of individuals. With reference to the constitutionally protected right to health (which is, however, always linked to the protection of the vital functions of the individual), this partially different outlook, in addition to a certain 'quality of life', would appear to suggest that the development and accomplishment of the individual is to be considered protected. The 'Aquilian'* protection for damage to the environment is thus strengthened.

Nevertheless, given the atypical character of the unlawful act under Italian legislation, it would perhaps not appear absolutely necessary to identify such an all-encompassing and complex subjective right (which is not easy to outline, even through an ambitious and systematic interpretation) in order to consider injury to the interests of individuals concerning the quality of the environment as being a proper matter for compensation.

It should not, however, be believed that as a result of these interesting developments the problem of the civil protection of the environment has been finally resolved. On the one hand, it has been maintained that the Aquilian protection of interests concerning environmental conservation is too unilateral and limited a remedy. On the other hand, the development in theory and practice in terms of environmental damage still has a long way to go. Besides, important developments could result from the wide debate under way on the subject of producer responsibility relating both to the possible criteria for allotting this responsibility and to the extension of the responsibility itself and the problem of cost–benefit analysis.

There is an ever-increasing tendency to represent the responsibility for unlawful damage caused by a business as a 'cost' for the company and therefore to make use of the regulations on civil liability as instruments for socialising the damage introduced into the lives of the population by industrial production, as well as to rationalise the 'costs' relating to the damage.

In this context, the problem of damage to the environment can represent an interesting application of the themes and concepts already developed

---

* 'Aquilian' protection is a legal concept in Italian law which has evolved from the old Roman Lex Aquilia, relating to wrongful damage to property.

in respect of other sectors, such as consumer protection. It should, however, be remembered that for protected interests such as health, Italian legislation also accords protection to social and collective interests, and not only to individual interests. Social damage suffered by a group as a whole does not coincide with the sum total of the damage to individuals, since it can consist of damage which can only be determined when the group is considered as a whole but only has minimal effect on a single individual and is in any case insufficient to reach the individual threshold for 'unjust damage'. Consequently, the cost–benefit analysis and associated theories such as the renowned 'business risk' (TRIMARCHI) are of use above all with regard to 'social damage' as such; certain limitations are revealed when they are used for individual damages which can lead to restrictions on compensation when the damage does not correspond with the typical risks which can be administered economically in certain business activity or when it represents a cost, the risk of which is acceptable to individuals because of the benefits arising from the production of certain goods or services.

Neither conclusion is always socially acceptable and, above all, it is not always in accordance with the supporting legislative structure.

With regard to the effectiveness of the legislation on civil liability, it is also necessary to note the doctrinal attempts made by authors of legal studies, to recognise, for the individual, inhibitory protection on the same basis as compensatory protection for the unlawful act. Modern tendencies (FRIGNANI, PROTOPISANI) identified the existence in Italian legislation of a general principle of protection through prohibition, by which jurisdictional actions to prevent harmful activities are tested out in a general context.

A further problem recently discussed is the use of the instruments so far described for the protection not only of individual interests, but also of collective or widespread interests. This is a question which finds wide application in the problem of environmental protection, also because individualistic remedies for environmental protection are of greater use, and a Civil Law approach is more justified (CORASANITI) in the collective dimension of protecting widespread interests in a community of individuals. Recent doctrinal contributions have turned their attention to models of other legal systems such as *class actions* and *public interest actions*, and have endeavoured to achieve similar results from the Italian legislative instruments. On the one hand an extension has been proposed of the use of the complex instruments of voluntary intervention in the process between other persons, and of the intervention by a third party to whom the case is common at the request of one of the parties, or by order from the judge (articles 105, 106 and 107 of the Civil Procedure Code). On the other hand, an attempt has been made to interpret less rigidly the provision on the subjective efficacy of the judgement (article 2909 of the Civil Code). It is nevertheless a question of somewhat limited remedies not comparable to the effects of a true *class action*.

But in fact the problem is somewhat different from one which can be resolved using instruments such as *class action*. This type of action only increases and multiplies the effect of a certain action brought to protect a right which in itself is already individually actionable, making it benefit anyone who finds himself in the same situation. However, in the light of all we have said, amongst other things with regard to social damage, or injury to collective interests, it is important to provide protection for interests which at individual level are not particularly protected (because they are 'simple' interests) since their injury does not reach the threshold of 'unjust damage', but which at a collective level are of particular significance, and even of primary significance. The 'widespread' character of these interests is, in this case, of particular importance since, at the same time, it is not possible to say that the interest of a single individual is involved, even only on a 'pro quota' basis. In this respect, the problem concerns possible associations and the individuals who can represent these interests in law. Court decisions have sometimes recognised and sometimes rejected the legality of acting to protect widespread interests of naturalist or conservation organisations (there is no specific regulation in Italy on this subject) by accepting the existence of an interest to act under article 100 of the Civil Procedure Code. In other cases, the initiative of associations or bodies has been facilitated by the fact that they are associations made up of individuals, whose work activity was directly damaged by the pollution (e.g. associations of fishermen). The problem is one of the nature, structure and importance which the association must have in order to represent effectively the interests of a collective group. Also in this instance the debate in progress in the neighbouring sector of consumer group protection is of assistance in the search for a solution (RUFFOLO). Indeed, the question of environmental protection, in its turn, can constitute a useful basis for clarification also in that parallel area of investigation, since it necessarily involves following a path different from that leading to 'consumerism', a course of action sometimes proposed in Italian legal practice which shows little understanding of the peculiar character of the Italian legislation and civil society.

This problem of widepsread interests relating to the environment, however, presents further facets (and also additional stimulus for a solution) if the public interest and administrative angles are considered. Not only, in fact, are these interests also actionable with regard to the public administration (within administrative jurisdiction) and can constitute a basis for non-jurisdictional action against acts by the public authorities, but they also play an increasingly important part in the formation process of the wishes of the public administration—that is to say, in the administrative procedures culminating in a discretionary act. Even though there is no general law in Italy on administrative procedure, nor any precise rules of debate between the various interests involved in the activity of the public adminstration, the idea is increasingly accepted that the discretionary administrative act is the result of the deliberations of more sectorial interests (GIANNINI) and the abstract concept of the public interest as an interest of the generality is weakened. Thus, the problem arises of introducing interests, even widespread, into this process and deliberation and, in this case, also of identifying the representative associations, groups and bodies.

There is, however, another aspect of the channelling of private interests, particularly collective and widespread, within the public administration: the fact that in some cases independent and repesentative public organisations of organised social groups (communes, mountain communities etc.) in turn make themselves the advocates of the groups they represent, taking part in civil or penal proceedings concerning matters which compromise their own territorial management functions.

As can be seen, the extent of remedies offered by what could be termed 'ordinary law' is not small, but the need for a more extensive and cohesive legislation is greatly felt. However, the existing special laws, by virtue of the fact that they introduce into the legislation elements of systematical interpretation (consider for instance article 1 of Law 615 of 1966 which would appear not only to protect the purity of the air as an asset in itself, but explicity refers to the need to prevent any direct or indirect prejudice to health), and by the general principles and teleological criteria they introduce, are useful to the interpreter also for solving the problems relating to the protection of private individuals (for whom those laws were not directly created).

Endeavours to achieve a better appreciation of the environmental problem and the real possibility of a response from Italian legislation have made little progress. However, there are examples of most useful activities in connection with the recognition and evaluation of data relating to the problem: at Parliamentary level, during the 5th, 6th and 7th legislatures, the Senate's Special Commission on Ecological Problems carried out with appreciable success experiments on a method for studying and analysing the problem which is particularly useful for obtaining an overall view of the situation (see 'Report on pollution from solid waste materials, the safeguarding of the wet-lands and noise pollution' published by the Commission during the 5th legislature, doc. XXV, No. 1).

Thus the activity undertaken within the area of the system of documentation and the preparation of data administered directly by the Court of Cassation on the part of GET (Ecology and Territorial Group) has been most important. This is organised by a group of magistrates who have been responsible not only for contributions to the study of the phenomenon but also a most valuable selection of legal precedents, inserted into a computer with terminals in all judicial offices, which thus constitutes a veritable mine of information on the evolution of the problem and the solutions adopted by the Courts. In addition, in Italy, in the context of public opinion, there is a recent but widespread understanding of the problems relating to the environment, to the land and to ecology. Echoes from the Stockholm Conference and the numerous initiatives encouraged by the United Nations Environment Programme and the Community directives on environmental matters have all contributed to this state of affairs.

Even though international understanding of the problem, encouraged by the above-mentioned factors, has increased the understanding of these subjects in Italy, it cannot be said that they have equally led to the adoption of coordinated, non-sectorial and comprehensive legislation. In addition, the recent delegation to the regions of wide legislative and administrative powers of management and control of the environment has certainly represented an evolution but at the same time has made it more difficult to formulate and follow certain lines of national environmental policy. The action of the individual regions, in fact, may shortly appear unhomogeneous and a temporary delay is expected in the application of suitable legislative and administrative instruments in regional areas where economic development is slow, whereas some of the more industrialised regions (such as Lombardy, Liguria, Tuscany and Lazio) are involved in a greater coordination between the protection of the environment and economic policy and planning.

There is no adequate coordination action on the part of the State organisation corresponding with this fragmentary regional action, in order to ensure a single cohesive policy. The law on water (Merli law and Merli bis) represents the only comprehensive national law in existence. There should be a framework law on environmental protection which, on the basis of the principles emerging from the Stockholm Conference, could formulate suitable legal instruments for the orientation of a national environment policy, channelling productive choices and management of environmental resources in a coordinated manner.

In fact, the environmental problem should be dealt with by means of a national policy of operative choices which should reconcile economic and social interests and will direct and coordinate regional policies.

The first problem which emerges is one of institutional reform: the present central structures are split up among the various ministries and, even though they are coordinated by Interministerial Committees, they can only ensure an environmental policy of maintenance and routine which tends more towards protection from pollution rather than prevention and management of the environment and resources.

For this purpose there are two possible solutions: a central organisation could be created in the form of a Ministry of the Environment—as in France—as an addition to the present ministers 'without portfolio' or 'with special duties'. Or, alternatively, a specialised agency could be created, made up of technical experts, responsible to the Prime Minister, operating at regional level through decentralised organisations, based on the model of the American EPA. In the first case the functions would be of a decision-making nature, combining in a single centre competence in environmental matters, a function which is at present divided up among too many ministries; in the second case, however, a compulsory advisory function would be involved with regard to the ministers of the sector and the Prime Minister.

In both situations the present Interministerial Committees could survive with coordinating functions in connecting competences, but there would be the undoubted advantage in having a central guiding force capable of taking initiatives or making proposals to be carried out in the various sectors.

The *Community directive* regarding the need to consider the environmental impact of every human activity and consequently the need to introduce regulations concerning the methods of preventive assessment has not yet been wholly followed in Italy. It is only at sectorial level and in a fragmentary manner that this method has been taken into consideration and partially implemented. It appears desirable and necessary to fill this gap, both for sectorial projects and for a general management plan for the environment. In particular, it would be essential, by means of analysis of the data collected, to predict the effects of a certain environmental policy on particular projects, and thus to operate programmed political and administrative choices in the light of a cost–benefit analysis in social terms (relationship between social costs and advantages of a certain use of environmental resources).

It is well known that many industrialised countries have introduced environmental impact assessments and analyses into their legislation as a regulatory procedure and fundamental component of their environmental protection policy (e.g. the USA, Canada, Federal Republic of Germany, France, New Zealand, Colombia, New Guinea). At international level this represents an objective on which the action of the United Nations Environment Programme is based (see the study and recommendations on environmental law by the intergovernmental working group of experts under UNEP 1977–80 to establish a worldwide directive on environmental policy), as well as the Community action (in particular the Action Programme of 1973).

It is therefore to be hoped that Italy, which shows a great understanding in the implementation of an environmental policy for the new international principles, will introduce into its own legislation measures aimed at ensuring the preliminary assessment of 'environmental impact' as an indispensable component for a comprehensive policy for the fundamental protection of the environment and for the promotion of a better quality of life.

9.2 ITALIAN LEGISLATION AND COMMUNITY DIRECTIVES

9.2.1

The problem regarding the carrying out of Community Directives in Italian legislation, and therefore the extent to which the Italian system is brought into line with the systems of other member States, is of particular importance in the environmental protection field because of the effects that potentially polluting activities carried out in one State can have on the national territory of another State.

There are three aspects to the problem: investigation into the ways in which the legal system is brought into line (when this has occurred); an assessment of how much has been achieved and how much remains to be done to implement the Community Directives; investigation into the compatibility of the Community Directives and national environmental legislation on the same subject, which have evolved independently.

9.2.2

Regarding the first aspect, the process of European legislative harmonisation has developed rapidly during recent years, creating a series of obligations for member States to bring their internal legislation into line, depending on the sources of legislation in force on the individual subjects. In Italy most of the Community Directives impinge upon relationships which are controlled by law and therefore involve the use of the State legislative process, with consequent delays in complying with Community obligations, both because of the complex and technical nature of the provisions to be accepted and because of the normal length of time necessary for the initiatives agreed upon by the various competent ministers and for approval by Parliament. Some laws relating to Directives have therefore only been passed after considerable delay and have only subsequently been implemented by regulations and ministerial decrees. Consequently, while a swifter method for implementing EEC Directives is being studied, the best method for rapidly issuing the necessary provisions at the present time is without a doubt by means of a delegation of legislative powers by Parliament to the Government.

For this reason, a draft law containing 'a delegation for the Government to issue implementing provisions for EEC Directives' (VIII Legislature— Chamber Act by the Deputies 1903/Senate Act 554) was presented to Parliament, but approval was delayed since it expired with the anticipated dissolution of the Chamber during the course of 1979, and was re-presented to begin the approval process over again. Under this law, Law 42 of 9 February 1982, the Government will issue decrees before 31 December 1981 containing the necessary provisions to implement a series of EEC Directives which are listed in the text of the law, including:

75/324 (Council) for the harmonisation of the legislation of the Member States relating to aerosols;

75/439 (Council) concerning the disposal of waste oils;

75/440 (Council) concerning the quality of surface water intended for the abstraction of drinking water in the member States;

75/442 (Council) relating to waste;

76/160 (Council) concerning the quality of bathing waters;

76/403 (Council) concerning the disposal of polychlorobiphenyls and polychlorotriphenyls;

76/769 (Council) concerning the harmonisation of legislative, regulatory and administrative provisions of the member States relating to restrictions on the marketing and use of certain dangerous substances and preparations;

77/312 (Council) concerning the biological screening of the population against the risk of lead poisoning;

78/319 (Council) relating to toxic and harmful wastes;

78/659 (Council) on the quality of fresh water which requires protection or improvement for the support of fish life.

A simplified mechanism for financing the expenditure required for the effective implementation of the provisions to be issued is also envisaged to ensure that there will not be further delays and thus that the causes of non-compliance vis-à-vis the commitment are removed.

9.2.3

With regard to the second aspect, during the course of this report it has been shown that the Community Directives on some matters have already been fully implemented, whereas implementing regulations have still to be issued on other subjects.

Thus, on the subject of noise pollution, Law 942 of 27 December 1973 relating to the 'inclusion in Italian legislation of the EEC Directive concerning the harmonisation of legislation in the member States relating to the type approval of motor vehicles and their trailers' left to ministerial decrees the regulation of subsequent technical data and EEC Directives on updating with scientific progress in the subject under discussion (decrees dated 5 August 1974 and 26 August 1977 issued by the Minister of Transport). Similarly, the Ministerial Decree of 5 May 1979 has implemented the EEC Directive on the noise of motorcycles.

Provisions are urgently required to deal with aspects of the problem which have already given rise in Italy to noteworthy judicial events: on the noise produced by aeroplanes (see 6.2.2) and the proposed Community Directive of 26 April 1976 in OJ C 126/76. In this sector a specific provision should be issued to enable the public administration at various levels of competence to implement the directives which are being discussed by the Community, such as the various proposed Directives of the Council regarding current generators (30 December 1975 in OJ C 54/76); tower cranes (30 December 1975 in OJ C 54/76); jack hammers and pneumatic concrete breakers (20 December 1974); construction machinery (77/113 of 19 December 1978 in OJ L 33/79).

Another sector in which Italian legislation is lacking with regard to Community Directives is waste disposal. As already mentioned (see 5.4), two draft laws have been presented which take into account the rules of the Community, especially those regarding the need to reclaim and recycle waste (see Couns. Dir. no. 75/422 of 15 July 1975 in OJ L 194/75) and the special regulations concerning certain categories of waste (at Community level see Couns. Dir. 78/319 of 20 March 1978 in OJ L 84/78 regarding toxic and harmful waste).

The subject of polluting products is one which has been brought partially into line with Community Directives (see Chapter 8). With regard to detergents, phytopharmaceuticals, insecticides etc. see 8.1.2. Where these products are concerned, the main problem is one of keeping abreast with all the updating and amendments to the Community provisions (e.g. the two Council Directives of 22 November 1973 in OJ L 347/73 on detergents and methods of controlling the biodegradability of anionic surface active agents—73/404 and 405; Council Directive of 23 November 1976 in OJ L 340/76 on pesticide residues in fruit and vegetables; Council Directive 75/324 of 20 May 1975 in OJ L 147/75 on aerosols; Council Directive 79/113 of 21 December 1978 in OJ L 33/79 on the prohibition on marketing and use of some plant protection agents). Italian provisions for other products are inadequate and do not correspond to Community Directives, as already illustrated (see 8.4 on used oils, the subject of Council Directive 75/439 of 16 June 1975 in OJ L 194/75). Some Community Directives relate to products used as fuel (e.g. Council Directive 78/611 of 29 June 1978 in OJ L 197/78, on the lead content of petrol); see section 2.3.3 regarding Italian provisions on the subject.

9.2.4

Air pollution and water pollution (inland and marine) are the sectors where the problem of compatibility with Community Directives is most acute.

In fact, the legislation on these subjects has largely developed independently from the need to harmonise European provisions, in spite of the fact that laws and other provisions have been passed to satisfy obligations at Community level (e.g. Law 437 of 3 June 1971 on pollution produced by combustion engine vehicles) and provisions have been included in recent laws to ensure compliance with international and Community obligations (e.g. article 11 of Law 319 of 1976 both in its old form and as amended by Law 650 of 1979).

The provisions concerning atmospheric pollution caused by combustion engine vehicles have been hasty and thus not very effective; furthermore, nothing has been done since 1971. Account has therefore not been taken of the amendments and updating carried out to Council Directive of 20 March 1970 (OJ L 76/70) by Council Decision of 28 May 1974 (OJ L 159/74) and by the Directive of the Commission of 30 November 1976 (OJ L 32/77) and Dir. Comm. of 14 July 1978 (in OJ L 223/78). The 1971 regulations for diesel engine vehicles (see 2.3) came into force before Council Directive 72/306 of 2 August 1972 (OJ L 190/72) was issued and it requires updating, not so much because of the abovementioned Directive, but because of the increasingly rigorous requirements of a regulation which is not limited to the moment of type approval and can impose severe sanctions (there is also Council Directive 77/537 of 28 June 1977 in OJ L 220/77 on agricultural diesel engines).

It should be pointed out that the Community Directives concerning the pollution of inland waters are more precisely aimed at achieving water quality objectives than is the case in Italian legislation, which concen-

THE ITALIAN JURIDICAL ORGANISATION

trates mainly on controlling discharges, as a result of the difficulties encountered in passing a law to control the use of water throughout its entire cycle (see 3.1 and 3.2). There is therefore a coordination problem regarding the Directives on the quality of water intended for use as drinking water (Council Directive 75/440 of 16 June 1975 in OJ L 194/75; Council Directive of 19 December 1978 in OJ L 5/79 etc.) and the quality of water suitable for fish life (Council Directive 78/659 of 18 July 1978 in OJ L 222/78).

With regard to the discharge of polluting substances into watercourses, Italian legislation is flexible and easily adaptable to the Community Directives (e.g. Council Directive 76/464 of 4 May 1976 in OJ L 129/76 on certain dangerous substances discharged into the aquatic environment of the Community). In this case, in fact, the most substantial difference between the Community Directive and Italian law concerns the permitted limits for substances in the discharge; but the tables annexed to Law 319 of 1976 and amended by Law 650 of 1979 can be revised by the Interministerial Committee provided by the law to bring them into line with the 'corresponding values defined by the appropriate Directives of the European Economic Community, if these are more stringent' (article 31 of Law 319 of 1976) (see 4.1.7).

Since Laws 319 of 1976 and 650 of 1979 also concern discharges into the sea, the same argument applies in this sector, particularly with regard to the need to coordinate these anti-pollution laws with the requirements of certain Community Directives which are centred on water quality objectives (see Council Directive 76/160 of 8 December 1975 regarding the quality of bathing waters).

In the nuclear energy sector, harmonisation is aided by the relative newness of the legislation, by the role played by the EURATOM Treaty within the legal system and by the monopoly exercised by ENEL, the National Energy Agency, over the installation of nuclear power stations. The Council Resolution of 3 March 1975 (OJ C 168/75) on energy and the environment underlined the role of energy saving as a means of protecting the environment. In Italian provisions for energy conservation there are more 'urgent' reasons linked to the energy crisis and the cost of petroleum.

Bibliography

AA.VV. *Inquinamento e salute umana*, ISVET, Roma 1970.
AA.VV. *Atti del Convegno "Uomo, natura e società" Frattocchie (Roma) 5–7 novembre 1971 – Roma*, Editori Riuntiti, 1974.
AA.VV. *Tecniche giuridiche e sviluppo della persona* (a cura di Lipari), Laterza, Bari, 1974.
AA.VV. *Intervento pubblico per la tutela dell'ambiente: ricognizione delle funzioni dello Stato e delle Regioni a statuto ordinario*, Giuffrè, Milano, 1975.
AA.VV. *Materiali per un corso di politica dell'ambiente* (a cura di Cannata), Giuffrè, Milano, 1975.
AA.VV. *Rassegna sulla tutela delle acque e dell'atomosfera* (a cura di Jannuzzi), Milano, 1975.
AA.VV. *Atti del Convegno "Le azioni a tutela di interessi collettivi"*, Pavia 11–12 giugno 1974, Padova, CEDAM, 1976.
AA.VV. *Atti del Convegno "Ecologia e disciplina del territorio"*, Pontremoli 29–31 maggio 1975, Milano, Giuffrè, 1976.
AA.VV. *Ecologia e disciplina del territorio*, Giuffrè, Milano, 1976.
AA.VV. *Atti del Convegno "Tutela pubblica dell'ambiente"*, Milano, 22–23 novembre 1974, Milano, Giuffrè, 1976.
AA.VV. *Tutela pubblica dell'ambiente*, Giuffrè, Milano, 1976.
AA.VV. *"Dalla lotta all'inquinamento alla tutela pubblica dell'ambiente"*, Milano-Bruzzano, 4–5 marzo 1977, Milano, Giuffrè, 1978.
AA.VV. *Ecologia e urbanistica* (a cura di Giacomini e Gorio), Milano, Giuffrè, 1978.
AA.VV. *Atti del Convegno "La responsabilità dell'impresa per i danni all'ambiente e ai consumatori"* Milano, 17–18 dicembre 1976, Milano, Giuffrè, 1978.
AA.VV. *La responsabilità dell'impresa per i danni all'ambiente e ai consumatori* (Atti del convegno di Milano 17–18 dicembre '76), Milano, Giuffrè, 1978.
A.A.V.V. *Atti del Convegno "Rilevanza e tutela degli interessi diffusi: modi e forme di individuazione e protezione degli interessi della collettività"*, Varenna, 22–24 settembre 1977, Milano, Giuffrè, 1978.
Agnetis, *Considerazioni sul problema ecologico*, in *Rass. Arma Carab.*, 1978, p. 27.
Alibrandi, Ferri, *I beni culturali e ambientali*, Milano, Giuffrè, 1978.
Alpa, Bessone, *Inquinamento atomsferico, "air codes" e forme "negoziate" di intervento pubblico*, in *Temi* 1974, p. 547 e *Pol. dir.* 1975, p. 136.

BIBLIOGRAPHY

Alpa, Bessone, *Tutela dell'ambiente, ruolo della giurisprudenza e direttive di "Common Law"*, in Riv. trim. dir. proc. civ., 1976, p. 227.
Amendola, *Tutela dell'ambiente: comparazione della legislazione italiana con quella di altri paesi, specialmente europei*, in Riv. dir. agr. 1975, I, p. 238.
Amendola, *La nuova legge sull'inquinamento delle acque*, Milano, Giuffrè, 1977.
Amendola, *Carenze di riforme e norme senza sanzioni in tema di inquinamento atmosferico da traffico veicolare*, Riv. circ. trasp. 1978, p. 368.
Amendola, *La legge n. 319 del 1976 e la completezza della domanda di autorizzazione allo scarico*, in Giur. merito 1978, p. 866.
Aquoli, *Autorizzazione ed inquinamento atmosferico*, Giur. merito, 1974, pp. 11–265.
Arvedi, *L'inquinamento atmosferico*, Inadel 1975, p. 713.

Bajno, Rampolla, Robecchi Majnarde, *Aspetti sostanziali, organizzativi e penali della tutela giuridica dagli inquinamenti*, in Riv. Dir. Min, 1971, p. 1.
Benvenuti, e A., *Inquinamento da piombo nell'ambiente circostante un colorificio ceramico*, Securitas 1973, p. 261.
Bessone, Roppo, *Strumenti di intervento amministrativo e "common law remedies" per una politica di tutela dell'ambiente*, in Pol. dir. 1975, p. 127 e in Pr. ped. 1974, II, p. 73.
Bessone, *Disciplina giuridica dei trasporti, sistema dei pubblici poteri e problemi di tutela dell'ambiente*, in Resp. civ. 1978, p. 345.
Burchi, *Evoluzione della riforma legislativa italiana in materia di protezione delle acque dagli inquinamenti*, in Riv. Trim. Dir. Pubb. 1975.
Burchi, *Profili finanziari nella legislazione italiana per la protezione delle acque dagli inquinamenti*, in Impr. amb. pubb. amm. 1976, I, p. 114.
Busetto G., *Inquinamento delle acque e agricoltura dopo la legge Merli*, in Giur. agr. it. 1976, p. 391.

Camera dei Deputati, *"Ambiente e informatica: problemi nuovi della società contemporanea"*, a cura del Servizio Studi legislazione e inchieste parlamentari, settembre 1974.
Carriero, *Tecnologia di controllo dell'inquinamento atmosferico da prodotti di produzione industriale*, Studi politica ambiente, p. 335.
Catelani, *Il codice delle leggi sull'inquinamento idro-atmosferico*, Firenze 1978.
Chirico, *Profili penalistici dell'inquinamento marino da polluzione di idrocarburi*, in Trasporti 1976, fasc. 8, p. 49.
Cicala, *Inquinamento atmosferico e legge penale*, Giur. it. 1974, II, p. 229.
Cicala, *La tutela dell'ambiente nel diritto amministrativo, penale e civile*, UTET, Torino 1976.
Cicala, *La tutela delle acque e del suolo dall'inquinamento nel passaggio fra "vecchia" e "nuova" disciplina*, note a T. Livorno 7 luglio 1976 in Giur. it. 1976, II, p. 623.
Cicala, *La normativa in materia ecologica nel quadro dei rapporti tra stato e regioni*, in Impresa ambiente pubb. amm. 1978, I, p. 98.
Colombini, *Inquinamento atmosferico ed educazione sanitaria*, Rass. amm. sanit. 1970, p. 1083.
Corasaniti, voce "Interessi diffusi", in *Dizionario di diritto privato* (a cura di IRTI), I, Giuffrè, Milano, 1979.

BIBLIOGRAPHY

De Marco, *Attività insalubri, inquinamento e giurisprudenza dei TAR*, in *Foro amm.* 1976, II, p. 513.

Di Cerbo, "Responsabilità penali degli imprenditori e dei dirigenti per le fonti e le cause d'inquinamento nelle fabbriche", in *Riv. infortuni* 1976, I, p. 495.

Di Giovine, "Getto pericoloso di cose ed inquinamento atmosferico", *Giur. Merito* 1973, II, p. 327.

Di Giovine, *Autorità sanitarie ed inquinamento delle acque* (nota a P. Vicenza 22 dicembre 1971), in *Corti Brescia*, Venezia, Trieste, 1974, p. 404.

Di Giovine, *Ecologia e disciplina del territorio*, in *Giur. agr. it.* 1975, p. 510.

Di Ronza, *Aspetti giuridici dell'inquinamento idrico*, Jovene, Napoli, 1975.

Giampietro, *I beni di proprietà collettiva nella nuova legislazione contro l'inquinamento*, in *Foro amm*, 1974, II, p. 453.

Giampietro, *Problemi di tetela ecologica*, in *Critica pen.* 1977, p. 195.

Giampietro, *Commento alla legge sull'inquinamento delle acque e del suolo*, Giuffrè Milano, 1978.

Giampietro, *Più concretezza e meno ottimismo nelle possibilità operative della legge "Merli"*, in *Giur. it.* 1978, IV, p. 81.

Giampietro, *Decentramento regionale e "legge Merli"*, in *Giur. it.*, 1978, IV, p. 14.

Giannini, *"Ambiente": saggio sui diversi suoi aspetti giuridici*, in *Riv. trim. dir. pubb.* 1973, p. 15.

Greco, Lazzaro, *La tutela delle acque dall'inquinamento*, Giuffrè, Milano 1977.

Ianni, *Il problema dell'inquinamento dell'aria da esalazioni di rifiuti solidi urbani e di polveri di cemento*, *Giur. agr.* 1974, p. 675.

Isvet, *Lineamenti per una politica di intervento pubblico contro l'inquinamento*, Collana ISPE, 1976.

Italia, *Problemi di diritto regionale sulla legge n. 319 del 1976 (Legge "Merli") e sul d.P.R. n. 616 del 1977*, in *Giur. it.*, 1978, IV, p. 120.

Kanitz, *Livelli attuali dell'inquinamento atmosferico da traffico motorizzato*, Roma, 1965, p. 83.

Liberti, *L'inquinamento atmosferico: considerazioni introduttive. Studi politica ambiente*, p. 289.

Locati, *Convergenza di potestà normative regionali e statali o potestà esclusiva dello Stato in materia di azione contro gli inquinamenti?*, in *Nuova rassegna*, 1975, p. 1221.

Lupo, "Una discutibile sentenza sul prelievo di campioni di scarichi in acque pubbliche" (nota a Cass. 22 ottobre 1977), in *Mass. pen.*, 1978, p. 311.

Malvici, *I compiti dei Comuni in materia di lotta contro l'inquinamento atmosferico*, *Foglio Inform.* 1973, p. 167.

Malvici, *Inquinamento atmosferico. Le norme applicabili e gli strumenti giuridici di intervento*, Foglio inform. 1973, p. 137.

Malvici, *La lotta agli inquinamenti – Inquinamento atmosferico – La inclusione dei Comuni in zone di controllo;* Foglio inform. 1973, p. 137.

Mammarella, *Inquinamento atmosferico in Italia* (Doc. ISVET n. 27), Roma 1970, p. 8.

Martino, *Ruolo delle regione nella tutela dell'ambiente*, in *Corti Br. Ve. tr.* 1974, p. 199.

Mazza, *Sui rapporti fra l'art.674 c.p. e l'art. 20 1. 13 luglio 1966, n. 615* (nota e Trib. Torino 18 marzo 1976), in *Giur. agr. it*, 1977, p. 164.

BIBLIOGRAPHY

Merli, *L'apporoccio italiano ai problemi ecologici, con particolare riguardo alle legislazioni sulle acque*, in Impr. amb. pubb. amm., 1976, I, p. 57.
Merusi, *Prospettive per una gestione pubblica dell'ambiente*, in Pol. dir. 1977, p. 57.
Micchiché, *Inquinamento da scarichi industriali*, in Nuova rass. 1977, p. 67.
Mineo, *Tutela ecologica delle acque – Alcuni profili di diritto amministrativo*, in Impresa amb. pub. amm., 1974 I, p. 84.
Mola, *Prevalenza della nuova disciplina sull'inquinamento atmosferico sulla vecchia disciplina di cui all'art. 216 t.u. leggi sanitarie del 1934*, Rass. Giur. ENEL, 1972, p. 948.
Morsillo, Esposito, Gotti-Porcinari, *L'inquinamento atmosferico*, Roma 1971.
Morsillo, *Inquinamento da allevamenti zootecnici e poteri del sindaco* (nota a P. Montejano, 30 genn. 1976), in Giur. agr. it. 1977, p. 175.
Pantano, *Legislazione e inquinamenti atmosferici*, ISLE, Roma, p. 27.
Paradiso, *Inquinamento delle acque interne e strumenti privatistici di tutela*, in Riv. Trim. dir. proc. civ., 1977, p. 1391.
Patti, voce "Ambiente (tutela civilistica)" in *Dizionario di diritto privato* (a cura di IRTI), I, Guiffrè, Milano, 1979.
Patti, *La Tutela civile dell'ambiente*, CEDAM, Padova 1980.
Pennacchia, *Inquinamento atmosferico e impianti termici civili. Studi politica ambiente*, p. 301.
Pisana, *L'amministrazione italiana e l'inquinamento delle acque*, in Cons. stato 1977, II, p. 50.
Pototsching, *Lotta all'inquinamento e pubblica amministrazione*, Dir. econ., 1971, p. 320.
Primicerio, *Ancora in tema di poteri legislativi ed amministrativi delle regioni ordinarie a statuto speciale: settore degli inquinamenti*, Dir. Sanit. 1971, 2/3, p. 3.
Rinando, *Inquinamento atmosferico. Profilo sostanziale ed organizzativo della normativa nelle industrie insalubri e della disciplina antismog*, Rass. Giur. ENEL, 1973, p. 375.
Robecchi Majnardi, *La Tutela degli inquinamenti nella jurisprudenza più recente*, in Riv. ing. sanitaria 1976, p. 84 ss.
Robecchi Majnardi, *Aspetti giuridici dell'inquinamento atmosferico*, in Foro pad. 1977, II, p. 29.
Rodella, D., *Problemi per l'applicazione della legge Merli per la tutela delle acque dall'inquinamento e le leggi regionali*, in Amm. It., 1976, p. 1316.
Romano, *L'ordinamento vigente in materia di tutela ecologica con riguardo alle reponsabilità degli amministratori*, in Amm. It. 1976, p. 1633 e 1977, p. 12.
Salvatore, *Tutela dell'ambiente dagli inquinamenti – Profili costituzionali*, in Cons. Stato, 1976, II, p. 467.
Salvi, *Le immissioni industriali*, Milano, 1979.
Santaniello, *L'evoluzione della legislazione in materia di tutela dell'aria e delle acque dell'inquinamento*, Riv. amm. 1970, p. 817, 818.
Santaniello, *L'intervento dei pubblici poteri per il disinquinamento atmosferico*, in Rass. amm. sanità 1970, p. 143.
Santaniello e altro, *Strumenti legislativi in materia di inquinamenti*, Riv. trim. scienza pol. amm. 1971, pp. 2, 13.

BIBLIOGRAPHY

Santaniello, *Profili giuridici del problema dell'inquinamento marino*, in *Impresa amb. pub. amm.*, 1974, I, p. 23.

Scarpulla, *L'inquinamento dell'atmosfera. Aspetti giuridici e tecnici*, *Giur. Agr.* 1974, p. 71.

Schinaia, *La gestione dell'ambiente tra stato e regioni*, in *Impresse ambiente pubb. amm.*, 1977, I. p. 475.

Schmidt di Friedberg, *Ecologia concreta; il caso Scarlino*, in *Impresa amb. pub. amm.*, 1974, I, 129.

Smitti, *La tutela delle acque dall'inquinamento – Note esplicativa sulla 1. 10 maggio 1976 (leggi Merli)*, Napoli, 1978.

Storey, *Inquinamento idrico e analisi economica del diritto*, in *Pol. dir.*, 1977, p. 527.

Strassoldo, *Sistema e ambiente – introduzione alla ecologia umana*, Milano 1977.

Tarantino, *La provincia ed i problemi dell'ecologia*, in *Nuova Rass.*, 1977, p. 389.

Tassara, De Ferrari, Andreoni, *Il controllo analitico delle acque inquinate*, Etas libri, Milano, 1975.

Trapanese, *Funzione dello Stato nella repressione degli inquinamenti*, in *Nuova Rass.*, 1974, p. 1892.

Treccani, *In tema di immissioni rumorose e di tutela del diritto alla salute*, in *Processi civ.*, 1978, p. 192.

Villa, *Il controllo in continuo sull'inquinamento ambientale – schemi di attuazione di gestione*, in *Difesa soc.*, 1978, fasc. 1, p. 54.

Visintini, *Immissioni tutela dell'ambiente*, in *Riv. trim. dir. proc. civ.*, 1976, p. 689.

Werner, voce, "Inquinamento"; Amendola, voce "La tutela giuridica dell'ambiente" in *Enciclop. Europea*, vol. VI, Garzanti, Milano 1978.

Classified Index*

The Constitution, Public Authorities, Special Interest Groups and Individuals

The national constitution	1.2.1
Sources of laws governing pollution control and remedies for damage caused by pollution	1.1.1
Government departments and agencies with supervisory, administrative or executive powers of pollution control	1.2
National, regional and local public authorities with powers of pollution control	1.2, 1.3
Independent advisory bodies with rights or duties under pollution control legislation	1.3.2
Special interest groups representing those who may be liable for pollution, or those concerned to prevent or reduce pollution	1.4
Standing to sue *(locus standi)* in legal proceedings for pollution	1.4.3

Air

Stationary Sources   2

Control by land use planning	2.1.2
Controls over plant and processes (including raw materials, e.g. fuels)	2.2, 2.2.5
Controls over treatment before discharge, and over manner of discharge (e.g. height of chimney)	2.2.3
Limits on emissions	2.2.4, 5
Enforcement, including monitoring and surveillance by or on behalf of the control authority	2.2.6
Ambient air quality standards	2.2.5
Rights of the individual	2.2.7

---

* References are to section numbers.

CLASSIFIED INDEX

Road Vehicles   2.3

Controls over design and construction	2.3.1
Requirements concerning maintenance	2.3.2
Controls over use, including time and place of use	2.3.2
Fuel which may be used or supplied for use	2.3.3
Enforcement, including monitoring and surveillance by or on behalf of the control authority	2.3.4
Rights of the individual	2.3.5 (see 1.4, 2.2.7)

Aircraft, Hovercraft and Ships   2.4

Inland Waters (Including Groundwaters)

Stationary Sources   3

Control by land use planning	3.1
Controls over processes, storage and other operations near inland waters (including waste disposal)	3.3
Controls over treatment before discharge and treatment plant	3.3
Controls over characteristics, quantities and rates of discharge	3.5
Controls over dumping in and on land near inland waters	3.5
Enforcement, including monitoring and surveillance by or on behalf of the control authority	3.6
Inland water quality standards	3.5
Rights of the individual	3.7 (see 1.4)

Seas: Pollution by Substances Other Than Oil

Coastal Waters   4

Definition of boundaries	4.1
Control by land use planning	4.1.1
Controls over processes, storage and other operations near coastal waters (including waste disposal)	4.1.3
Controls over treatment before discharge	4.1.3
Controls over characteristics, quantities and rates of discharge	4.1.3
Monitoring to be done by discharger	4.1.3
Enforcement, including monitoring and surveillance by or on behalf of the control authority	4.1.5
Coastal water quality standards	4.1.7
Rights of the individual	4.1.6 (see 1.4)

CLASSIFIED INDEX

Controls over Dumping from Ships and Aircraft   4.2

Geographical extent of jurisdiction	4.2.1
Monitoring	4.2.1
Enforcement	4.2.1

Control of Pollution Caused by Exploitation of the Sea and Sea Bed   4.2.2

Geographical extent of jurisdiction	4.2.2
Monitoring	4.2.2
Enforcement	4.2.2

Seas: Pollution by Oil

Controls over Discharges from Ships   4.2.3

Discharges of oil	4.1.4, 4.2.3
Enforcement, including surveillance by or on behalf of the control authority	4.2.5
Civil liabilities	4.2.5

Controls over Shore Installations and Port Activities   4.1.4, 4.1.5

Controls over Offshore Installations and Operations   4.2.2, 4.2.5

Geographic extent of jurisdiction	4.2.2
Licensing	4.2.2
Construction, equipment, safety zones	4.2.2
Manning	4.2.2
Discharges	4.2.2
Loading and unloading	4.2.2

Contingency Plans for Oil Pollution Incidents   4.2.4

Discharges to Sewers

Prohibitions	3.4
Controls over characteristics, quantities and rates of discharge to sewers	3.5
Requirements for treatment of collected sewage	3.5
Controls over quality of discharges from sewers or sewage treatment plants	3.5
Enforcement, including monitoring and surveillance by the control authority	3.6
Civil liabilities	3.6

CLASSIFIED INDEX

Disposal of Waste on Land

Controls by land use planning, including licensing of sites for treatment and disposal	5.1.1
Controls over treatment before disposal, including recycling and reclamation	5.1, 5.1.2, 5.1.3
Controls over methods of disposal	5.1.2
Enforcement, including surveillance by or on behalf of the control authority	5.2, 5.4
Rights of the individual	5.3 (see 1.4)

Noise and Vibration

Stationary Sources   6.1

Controls by land use planning, including licensing	6.1.1
Controls over design and construction of noise generating plant and equipment	6.1.2
Restrictions on emissions	6.1.3
Enforcement, including monitoring and surveillance by or on behalf of the control authority	6.1.4
Ambient noise standards	6.1.4
legal requirements, objectives or guidelines	6.1.4
evidence of achievement of standards (may be by reference to sources of information only)	6.1.4
Rights of the individual	6.1.5 (see 1.2.3.1 and 1.4)

Road Vehicles   6.2/6.2.1

Controls over design and construction	6.2.1.1
Emission standards	6.2.1.1
Requirements as to maintenance	6.2.1.1
Controls over use, including times and places of use	6.2.1.1
Enforcement, including surveillance by or on behalf of the control authority	6.2.1.2
Rights of the individual	(see 1.4)

Aircraft   6.2.2

Controls over design and construction	6.2.2.1
Emission standards	6.2.2.1
Requirements as to maintenance	6.2.2.1
Controls over use, including times and places of use	6.2.2.1
Enforcement, including surveillance by or on behalf of the control authority	6.2.2.2
Rights of the individual	(see 1.4)

Nuclear Energy

Nuclear Installations/Legislation on Nuclear Matters   7.1.1

Controls by land use planning, including licensing	7.2/7.2.1
Controls over design and construction	7.2.2
Controls over maintenance and operation	7.2.3
Controls over accumulation of discharge of wastes	7.2.4
Obligations on the part of the installation operator	7.2.5
Enforcement, including monitoring and surveillance by or on behalf of the control authority	7.2.6
Rights of the individual	7.2.7

Radioactive Substances   7.3

Controls over storage and use	7.3.1
Controls over packaging and transport	7.3.2
Controls over accumulation and disposal of wastes	7.3.3
Monitoring to be done by the discharger	7.3.4
Legal standards, objectives or guidelines for levels of radioactivity in the environment	7.3.5
Enforcement, including monitoring and surveillance by or on behalf of the control authority	7.3.1/7.3.5
Rights of the individual	7.3.6

Controls Over Products

Pesticides	8.2
Detergents	8.1
Fuels/mineral oils	8.4
Foods	8.3
Toxic gases	8.5

Environmental Impact Assessment

Requirements for environmental impact assessment	9.1
Italian legislation and Community Directives	9.2

I.e. controls over products introduced for the purpose of protecting the external environment.